D0087541

TRAVELING AT
THE SPEED OF THOUGHT

Traveling at the Speed of Thought

EINSTEIN AND THE QUEST FOR GRAVITATIONAL WAVES

Daniel Kennefick

Princeton University Press

Princeton & Oxford

Copyright © 2007 by Princeton University Press

Published by Princeton University Press, 41 William Street,
Princeton, New Jersey 08540

In the United Kingdom: Princeton University Press, 3 Market Place,
Woodstock, Oxfordshire OX20 1SY

All Rights Reserved

ISBN-13: 978-0-691-11727-0
ISBN-10: 0-691-11727-6

Library of Congress Control Number: 2006938366

British Library Cataloging-in-Publication Data is available

This book has been composed in Adobe Garamond

Printed on acid-free paper. ∞

pup.princeton.edu

Printed in the United States of America

1 3 5 7 9 10 8 6 4 2

Selection from Richard Feynman's letter printed with
permission by Melanie Jackson Agency, LLC.

TO MY PARENTS

Dan and Maura Kennefick

CONTENTS

ILLUSTRATIONS

ACKNOWLEDGMENTS

The research that led to this book was the idea of my physics advisor at Caltech, Kip Thorne, who had been involved in some of the controversies discussed in it and saw that it was a story that someone should try to tell. I was fortunate in having an advisor who was so open to the idea that I should combine studies in physics with work in the history of science. I was doubly fortunate in finding a history advisor as able to facilitate my novice efforts in so many ways as Diana Buchwald, now the general editor of the Einstein Papers Project, engaged in editing the Collected Papers of Albert Einstein. After finishing my thesis work, I benefited greatly from a two-year period spent collaborating with Harry Collins, author of *Gravity's Shadow*, a history of experimental gravity wave physics, at Cardiff University.

Diana made sure I got much excellent advice at an early stage from the first editor of Einstein's papers, John Stachel. Another physicist turned historian, Peter Havas, who like Kip was centrally involved in much of the history I was studying, was initially skeptical that I could objectively tell this story, given my background and training. Yet he never let this prevent him from helping me in every way he could, in person and by correspondence.

I owe a particular debt of gratitude to the many physicists who agreed to be interviewed by me and would like to thank them all. At the end of the book I have provided a table of the interviews and when they took place. They all informed the text considerably, even when they are not quoted directly. Funding to conduct these interviews was provided by the National Science Foundation through a Doctoral Dissertation Improvement Grant.

A number of people kindly read the manuscript and offered much helpful advice, in particular Martin Krieger and Eric Poisson, as well as Tilman Sauer, Michel Janssen, and David Kaiser.

I would like to thank the Albert Einstein Archives at the Hebrew University of Jerusalem for permission to quote from the correspondence of Albert Einstein. The California Institute of Technology has kindly granted me permission to quote from correspondence of Howard Percy Robertson and from a letter of Richard Feynman's in their possession, for which I am also grateful. They have also kindly granted permission for the whole of Robertson's referee report of a paper by Einstein and Nathan Rosen to be reproduced in transcript in appendix A. I would also like to thank Michelle Feynman for her kind permission to quote from her father's correspondence. The mathematician John Tate, Jr, graciously granted permission to quote from his father's letters, for which I thank him, and also Nathan Rosen's sons, Joe and David Rosen, who granted permission to quote from their father's letters to Einstein.

My wife, Julia Kennefick, also critiqued the manuscript and provided constant encouragement and advice from the earliest days of this research up to the present. Finally I would like to thank my parents, Dan and Maura Kennefick, without whose love and encouragement I would never have properly begun this work and to whom I dedicate this book.

TRAVELING AT
THE SPEED OF THOUGHT

1

The Gravitational Wave Analogy

In the early years of the twenty-first century, several large detectors designed to be the first "observatories" of gravitational waves went, one by one, into operation. These detectors trace their ancestry to around 1960, when Joseph Weber, an American physicist working at the University of Maryland, first began the experimental effort to detect gravitational waves. Until 1969 the "field" of gravitational wave detection consisted of Weber and his students, but when he claimed to have detected gravitational waves (Weber 1969), others (some of whom had previously considered working on this subject) began to build their own devices. It proved to be a false dawn. In a highly controversial episode lasting several years, the new detector groups all failed to replicate Weber's results with their instruments (Collins 2004). Nevertheless, despite this controversial and turbulent start, most of these groups continued in the field, persisting through decades of hard effort and many different instruments. Today it is widely expected that the first direct detection of gravitational waves will take place within the decade.

As an illustration of how likely the detection of gravitational waves was thought to be in earlier years, one of the most dedicated boosters of the effort, Kip Thorne, had no difficulty in 1981 in finding a taker for a wager that gravitational waves would be detected by the end of the last century. The wager was made with the astronomer Jeremiah Ostriker, one of the better-known critics of the large detectors then being proposed. Thorne was one of the chief movers behind the largest of the new detector projects, the half-billion-dollar Laser Interferometer Gravitational Wave Observatory, or LIGO. He lost the bet, of course. One can see the record of it posted in the west corridor of the Bridge Building at the California Institute of

Figure 1.1. The LIGO facility at Hanford, Washington, one of two separate detectors in the LIGO system. This Laser Interferometer Gravitational Wave observatory consists primarily of two 4-km beam tubes down which lasers are fired and bounced back and forth by high-quality mirrors. Any changes in the time taken for the light to travel along the tubes is possible evidence for the passage of gravitational waves. One of the two beam tubes can be seen receding into the distance. (Courtesy LIGO Laboratory)

Technology (Caltech), outside Thorne's office. It stands beside more than half a dozen other such wagers between Thorne and his colleagues (the most famous of Thorne's wagering friends is Stephen Hawking), most of which Thorne has won. On it is written his note of concession, "I underestimated the time required to make LIGO a reality." It is actually more remarkable that he and others managed to make LIGO a reality at all, especially when one considers the controversial history of the field, for that controversy was not only on the experimental side. The theory of gravitational waves has an even longer history of disputes, false dawns, and setbacks. At one time or another, many theorists doubted whether such waves existed at all. Albert Einstein himself, who founded the theory of gravitational waves in 1916, numbered himself among the doubters on at least two occasions. How that controversy came to be replaced by the certainty and conviction necessary to motivate the great projects of today is the principal story of this book.

The ambivalent status of gravitational waves within physics at the time of the Weber controversy (the early 1970s) is expressed in the following comments relating to Weber's claims from the National Academy of Sciences report *Astronomy and Astrophysics for the 1970s*. This report was written in 1971 as a guide to U.S. government funding agencies, in particular the National Science Foundation, for the decade ahead:

> The detection of gravitational waves bears directly on the question of whether there is any such thing as a "gravitational field," which can act as an independent entity. All actively pursued gravity theories deal with the concept of a gravitational field, so the mere existence of gravitational waves does not exclude any of these theories. Thus this fundamental field hypothesis has been generally accepted without observational support. Such credulity among scientists occurs only in relation to the deepest and most fundamental hypotheses for which they lack the facility to think differently in a comparably detailed and consistent way. In the nineteenth century a similar attitude led to a general acceptance of the ether and atoms decades before the experiments that abolished the ether and confirmed the atom.
>
> The basic style of all physics so far in the twentieth century has been set by the field concept, which arose in electromagnetic theory to replace the vanished ether. This idea has been so overwhelmingly convincing, when tested in experimental and industrial applications, that scientists have tried to package every other known fundamental domain of physics in the same mold. Field theory is incontrovertibly successful in the case of the electromagnetic field. Application of field concepts to particle physics has been successful in many respects, but there are still many unresolved problems. Confirmation of the gravitational wave experiments would show that this concept is suitable for at least one of the other areas—that of gravitational phenomena—in which it is customarily employed. (pp. 282–283)

Thus belief in the existence of gravitational waves, while unsupported by any physical evidence as of the early 1970s, nevertheless prevailed among most physicists because they could see no alternative to modeling the gravitational force other than by analogy with the field theory that described electromagnetic theory. In 1916, shortly after his discovery of the field equations of general relativity, Einstein became the first to describe gravitational waves within a complete field theory of gravity. Beginning with his approach, relativity theorists looked to various analogies with the electromagnetic field as they attempted to construct a theory of gravitational waves in the absence of experimental evidence. Along the way, disputes arose over the theoretical description of this phenomenon and even over whether the theory predicted its existence at all. Given this skepticism, it is not surprising to find that some relativists regarded the analogy underpinning faith in the orthodox picture of gravitational waves as inadequate. It is interesting to examine the views of those whose skepticism made them defy the consensus understanding of a hypothesis without which modern physicists lacked "the facility to think . . . in a . . . detailed and consistent way."

How is it that an idea can be universally, or at least "generally," held in science "without observational support?" To most ears, whether they belong to scientists or lay people, this does not sound terribly scientific. However, such a state of affairs is actually not uncommon in science, because it does happen that an extremely useful concept (e.g., that of a "field" of force), which is designed to provide an underlying explanation for observed phenomena, may arise without itself necessarily being detected or detectable. The case of gravitational waves is admittedly a little more unusual. In this case, we have a phenomenon that is suggested by the field concept as something whose effects on matter should be detectable. However, the phenomenon has never been detected and has been without even indirect evidence for decades, yet widespread belief in its existence persists. Some rather powerful motivation must lie behind this belief, and if we look for it, we find that the force behind this extraordinarily tenacious scientific belief is that of analogy. Specifically, the analogy in question is between the gravitational field, which underlies our modern theory of gravity (general relativity), and the electromagnetic field, which originated in the work of Michael Faraday and James Clerk Maxwell in the nineteenth century and which is the centerpiece of modern physics. The most dramatic prediction and confirmation of Maxwell's theory was the existence of radio waves, which are part of a spectrum of electromagnetic radiation that actually includes light itself. The basic analogy here is that if gravity is described by a field theory, should it not also have waves which play the fundamental role in that theory as electromagnetic waves do in the theory of Maxwell and his successors?

The history of the field concept is itself a controversial one, dominated by arguments from analogy. Once the idea of electromagnetic radiation became widely accepted, nineteenth-century physicists naturally looked to analogies with other wave phenomena, like sound, which suggested that waves require a medium (such as air in the case of sound) to propagate in. No medium means no wave. The field concept became associated in the nineteenth century with the idea of the luminiferous ether, which was an invisible substance with bizarre properties that pervaded all of space and was the medium or carrier for the electromagnetic field. The old ether theory was discarded completely early in the twentieth century, and nowadays, insofar as we say that electromagnetic waves have a medium at all, we say that that medium is the electromagnetic field, an entity which is not even a part of the material world, although it is, of course, generated by particles that make up the material world. The electromagnetic field is generated only by particles which carry electric charge, but since energy has mass, all particles that exist (i.e., have energy) in the material world generate a gravitational field. Keep in mind, of course, that particles are themselves idealizations designed to help physicists model matter in their equations. In some sense we do not directly experience fields as real entities at all but instead observe their influence on the matter that surrounds us. Thus an electromagnetic wave exists for us only in so far as we have some device at hand that can absorb energy from it as it passes by.

Now how is it that such highly abstract ideas as fields of force have come to play so important a role in physics? It happened in stages, with the level of abstraction increasing at each step. This development went hand in hand with the increasingly dominant role of mathematics in physics. At one time, for instance before Newton, it used to be thought that physics, which was primarily involved with explaining the *qualities* of physical things, was not a very suitable discipline for the use of mathematics. One characteristic of modern physics since Newton has been an escalating mathematization of the subject. Indeed, Einstein's introduction of general relativity played a major role in this process, as did Newton's introduction of his gravitational theory in the *Principia*. A very important agent of this increasing abstraction has been the creative use of analogies. For instance, in ancient times Greek philosophers and Roman engineers proposed that sound was a kind of wave traveling through the air, drawing an analogy with the motion of waves on the surface of bodies of water. Via the physics of Aristotle, this concept passed into the physics of the Middle Ages. At the time of Newton, through the work of his contemporary, Christiaan Huygens, and again in the nineteenth century, it was proposed that light was also a wave phenomenon, based on an analogy with the propagation of sound. At this stage the analogy was already becoming further removed from the immediate physical source of the metaphor, because advocates of the wave theory of light were more apt to make a comparison with sound rather than directly with water waves. The disconnection from direct physical experience increased greatly in the second half of the nineteenth century with the development of the Maxwellian theory of electromagnetic radiation, owing to the work of Maxwell, Heinrich Hertz, and others. A further stage of abstraction was added by the early relativity theorists in the first decades of the twentieth century, when they hypothesized that gravitational waves might exist in a field theory of gravitation. They based their analogy on the case of electromagnetic waves, already several stages removed from the kind of wave that we can actually see operating on the surface of water. Also, in this case, the extension of the analogy was being used to *predict* rather than to explain the existence of a new phenomenon. Some of the great drivers of change in twentieth-century physics were the discoveries of new forms of radiation and new particles, yet gravitational waves were to remain in a kind of limbo, predicted but not observed, for most of the century.

It is worth mentioning that the analogy with electromagnetism was a powerful tool for Einstein in his discovery of the field equations of general relativity. The route Einstein took to create this theory was a long and difficult one and has been extensively studied in recent years.[1]

Much has been written about analogies and their use in science (see especially Hesse 1966), but it is clear that there is no straightforward definition of what a scientific analogy is. Most analogies, like the one between gravitational and electromagnetic waves, could also be described as *models*. It is obvious that when physicists talk about gravitational waves being analogous to electromagnetic waves, they are thinking of the latter as a model for the former. The use of models is highly characteristic of physics, and in preferring the term *analogy* in this case, I am doing so because that is the term most often used by the physicists who work in this field. In addition I think it helps to clarify exactly what kind of model we are talking about. There are models that physicists hold to be actually true depictions of the thing being modeled, for instance, the kinetic model of gases, which visualizes them as consisting of many tiny molecules. Nowadays physicists believe that this is exactly what gases consist of. A model may also be a kind of construction of many parts, each of them imaginary, but whose whole forms a functionally equivalent depiction of the thing modeled (what we have in mind when we speak of "model-building"). An example of this kind would include Maxwell's attempts to model the luminiferous ether as a mechanical system of gears and cogs. In our case physicists do not believe that gravitational waves really are electromagnetic waves (though when the idea first emerged this seemed a real possibility), nor are they constructing a model from simple building blocks. They are saying that gravitational waves behave like electromagnetic waves and that there is often a point-for-point comparison to be made between the equations which govern each. This makes the word *analogy* a very apt term to describe what is going on, since when we talk about analogy we often understand by the term a set of correspondences between two systems, such that features of one system correspond to certain features of the other system. The analogy that physicists studying gravitational wave refer to is a rather formal, mathematically based analogy, which establishes that there are correspondences between the equations governing gravitational waves—and the quantities appearing in them—and the equations governing electromagnetic radiation. However, there are other analogies of interest to us that are more descriptive and informal in nature, such as claims that gravitational waves can be thought of as "ripples in the curvature of spacetime," evoking water waves on the surface of a pond. To be clear I will use the term *metaphor* to describe this descriptive kind of analogy and try to reserve the term *analogy* for the more formal one that lies at the heart of my argument.

For our purposes we will focus on two main uses of analogy, the first of which is as a heuristic, or finding, device. In this case, one is not necessarily committed to making every point of the analogy correspond, because one is not supposing that the two entities are really the same thing. Thomas Kuhn has discussed the importance of this kind of analogy in the practice of physics, emphasizing what a critical role analogic thinking plays in enabling physicists to make the widest possible use of the tools available to them. Although physicists come to a new problem with a large repertoire of mathematical techniques for solving problems, it will not be immediately clear how best these techniques can be applied to the problem at hand (Kuhn 1977, pp. 306–7).[2] Kuhn argued convincingly that looking for analogies between the new problem and those solved successfully in the past is the chief way in which a physicist will try to deal with an unfamiliar topic: "Once that likeness or analogy [between the new problem and the old] has been seen, only manipulative difficulties remain" (Kuhn 1977, p. 305). The beauty of finding such an analogy is that it enables the physicists to unleash their arsenal of techniques, hard won through experience in other subjects, onto a new problem. But since, as Kuhn emphasized, the sort of analogy that is being exploited is by no means a strict set of correspondences, there is plenty of room for argument about the validity of the analogy. As another philosopher of science has said, "Arguments from analogy may be fruitful, but they are always invalid" (Mario Bunge; quoted in North 1981, p. 135). There is plenty of scope for being wrong in this business. In cases where the dependence on analogy persists for a (perhaps unusually) long period, we will not be surprised to find that it is fertile ground not only for discovery, but also for controversy.

Now an analogy may also be proposed in a situation where it is suspected that a real underlying structural connection exists between the two entities being compared (as in the case of the kinetic model of gases). In this case the analogy may be viewed as the first step on the road to unifying two areas of physics. Thus from the beginning, gravitational waves were seen as an element of a possible unified theory of electromagnetism and gravity. This made the argument from analogy appear much more compelling to some physicists. In addition there was another argument, based on special relativity, that was derived from the fact that the principle of relativity (which guided the development of special relativity) presumed that no signals could travel faster than light. If gravity proved an exception to this rule, then this *might* threaten the underpinnings of this important theory. Therefore the analogy in this case acquired great force because it appeared that if gravity did not behave analogously to electromagnetism, at least in regard to the speed at which the field was propogated, then it would entail great problems for Maxwell's theory, which Einstein had only with difficulty reconciled to relativity theory.

The analogy on which wave theories are based naturally gave rise, as already mentioned, to attempts to identify a medium for phenomena like light and gravitational waves. When we look at waves in water we realize that the wave, though we can see it as a thing which moves and has its own reality, is at the same time nothing more than a disturbance in the water. It exists within a medium through which it moves or propagates. The medium may move as part of the disturbance, but it is not the medium which is propagating, or traveling, with the wave. The wave that washes against the California shore is not made up of water molecules freshly arrived from Japan. The water is local; it is only the disturbance that has traveled across the sea, being handed on, as it were, from one part of the water to another as it goes. Similarly, sound is a disturbance in the air, through which it propagates. Sound is not like wind; it doesn't move masses of air about but instead travels through it. What, then, is the medium that transmits or propagates electromagnetic and gravitational waves? The attempts to visualize the medium of electromagnetism in the nineteenth century as the luminiferous ether came to an ignominious end around the turn of the century amid failed attempts to detect any evidence for an effect on the behavior of light as it moved through the ether. If light had a medium, then its movement through the latter should surely affect its apparent speed, just as would be the case with waves in water. The famous Michelson-Morley experiment (among many other experiments) failed to detect any variation in the speed of light depending on whether it was moving into or along with the putative ether wind created by the motion of the earth. The luminiferous ether theory encountered increasing obstacles as the theory of electromagnetism developed, and Einstein dealt it a mortal blow with his special theory of relativity in 1905, which took as fundamental assumptions that the motion of all material bodies is relative, whereas light has the same velocity for all observers. He argued that without such an assumption the principle of relativity could not be consistently applied to all areas of physics. As this standpoint became the accepted one, it discouraged all attempts to treat the medium of light and other electromagnetic waves as a physical thing. As we shall see, the medium of gravitational waves, as presently conceived, is spacetime itself. The once abstract entities space and time take on a very active role in Einstein's theory. This entire process went along with an increasing formalism, which saw a retreat from the kind of highly visual modeling that characterized Maxwell's approach to the subject and a turn towards the feeling that it was only mathematics which mattered. In Hertz's famous phrase, "Maxwell's theory is Maxwell's system of equations." Thus we have the twentieth-century attitude in which physical models may not be entirely real, but equations certainly are, as expressed by the T-shirt, popular with scientists everywhere, which reads, "And God said, 'Maxwell's equations' and there was light." This outlook, which to some extent reflects Einstein's approach to physics in his later years, is characteristic of the physics of the last century.

Now we can discuss how all this talk of analogies is related to the theoretical "discovery" of gravitational waves. Let us give a brief summary of how it came about. First, we have Maxwell's theory itself, which took a force known to obey an inverse-square force law and rewrote it as a new kind of theory, a field theory, in which an associated wave phenomenon played a central role. This must have naturally suggested a simple and not very compelling analogy that probably influenced a few people prior to 1900. But afterwards, during the period of relativity theory, in which electromagnetic waves and their speed of propagation were seen to play an absolutely central role in physics in general, the notion arose that surely gravity could not be left in its old Newtonian form. Since the speed of light now appeared as an absolute upper limit to the speed with which signals could travel, how could the gravitational force make itself felt at great distances instantaneously, as the traditional theory demanded? That the speed of light was an upper limit to the speed with which any signal could be transmitted was of central importance to Einstein's demonstration that the new relativity transformations of the Maxwell-Lorentz equations also applied to the everyday science of motion, kinematics. This *upper limit* was at the foundations of the new theory of relativity. It seemed that the force mediated by the gravitational field must propagate at a finite speed (so that we feel the Sun's gravitational pull on us only after the effect has had a chance to cross the space between us), and this impression naturally suggested the idea of an associated wave. The phrase *gravitational wave* (*onde gravifique*) was first coined by Henri Poincaré in 1908 in this milieu. The idea had clearly become highly suggestible, and once Einstein had found his set of equations for the gravitational field in 1915, he immediately turned to describe gravitational waves by a method designed to make his equations look and behave as much like Maxwell's as possible. Thus the equations were the reality, and if the equations predicted the waves that the more metaphorical and less formal analogy suggested, then surely they must exist. The only remaining problem was that it was quickly recognized that the waves would be very hard, if not impossible, to detect on account of being very weak. Therefore an element of physical theory that seemed as if it must exist was nevertheless compelled to remain in the limbo of the unverifiable, as far as experiment or observation was concerned. But of greater relevance for most of our story was that the equations were far from telling all they knew. Because of their great complexity, Einstein had been obliged to make use of an approximate method that was open to many criticisms. Thus over the years, researchers returned time and again to these alluring creatures, with a degree of optimism or skepticism that varied according to the physicist's personality. Einstein was far from saying the final word on the subject, but he had sown the seeds of a great controversy. He himself would later help produce the spark that would set the whole field ablaze.

It is worth mentioning at this point why the shorter term *gravity waves* was not used until recently to describe this phenomenon. The reason is that this term had already been coined to describe certain types of waves in water whose motion is determined by the water's own weight, or gravity, rather than by the surface tension, which governs what we call ripples. Gravity waves are the longer wavelength waves with the characteristic curving peaks that we are accustomed to seeing on a trip to the beach. This use of the term *gravity waves*, which has been around since the nineteenth century, has been only very slowly displaced by its current use as a handier term for gravitational waves. Obviously, the longer phrase is something of a mouthful, but for as long as gravitational waves remained the esoteric subject of theory only, there was little reason to shift the meaning of *gravity waves*. However, as the subject of fluid wave mechanics has declined somewhat in importance (probably because of the peak of perfection it reached concerning phenomena such as waves in water), and as gravitational wave studies have become increasingly important in the last thirty years, the phrase *gravity waves* has come to be generally used to refer to gravitational waves, though I will prefer the latter term.

It is the rippling waves characteristic of relatively short wavelength water waves that provide the usual kind of visual analogy with gravitational waves. Although, as we have discussed, there are several layers of abstraction lying between our visual impression of water waves and the idea of gravitational waves, physicists often find it useful to reach back to the original source of the metaphor when attempting to describe gravitational waves for a lay or student audience. Probably everyone has heard the expression that gravitational waves are "ripples in the curvature of spacetime." Just as water ripples appear like smooth variations of curvature on a two-dimensional sheet, we expect gravitational waves to consist of curvature in the spacetime sheet (which is, of course, four-dimensional). Thus although the formal analogy with electromagnetic waves has played a much bigger role in the development of the theory of gravitational waves, the ripples metaphor, with its far greater visual appeal, has come to prominence in recent times in popular exposition. Interestingly, it was not used a great deal in popular accounts before about 1970. It is almost as if physicists were careful about the use of strong metaphors until they themselves had attained a greater understanding of gravitational radiation. Before this time, even when addressing a lay audience, they preferred to stay fairly close to the electromagnetic analogy. As their confidence in their mastery of this new and purely theoretical phenomenon grew through the 1960s and early 1970s, they ventured into more colorful explanations. It is perhaps no coincidence that this is also the period in which the first real hopes of experimental detection emerged. It was only when gravitational waves drew closer to the real world of observations that they could be described in terms of more concrete analogies with everyday experience.

I will use the term *skeptic*, as it was used by many relativists in this context, to describe those theorists who either doubted the existence of gravitational waves altogether or, more commonly, thought they would not be emitted by freely falling gravitational systems, such as binary stars, which today are considered the sources most likely to be observed by the new detectors. Although this use stretches the term to cover dozens of theorists over many decades from the 1920s to the 1980s, there is one attitude that seems common to each of them: their skepticism of the analogy with electromagnetism, which they regarded, for various reasons, as an inappropriate or misleading model on which to construct a theory of gravitational radiation. They were aware of the compelling force that this analogy had for many of their peers, as one leading skeptic, Leopold Infeld, made clear in his autobiography:

> It is therefore apparent that the existence of gravitational waves can be deduced from general relativity just as the existence of electromagnetic waves can be deduced from Maxwell's theory. Every physicist who has ever studied the theory of relativity is convinced on this point. (Infeld 1941, p. 261)

Nevertheless, although most of the skeptics accepted that the field concept was appropriate for gravitation, they were wary of assuming too many parallels between the gravitational field and the electromagnetic field.

The skeptics did not ignore the analogy; in fact, they often addressed it in their papers at greater length than did nonskeptics, but they did so critically. It also guided their thinking on the radiation problem, but in a different way from the use made of it by nonskeptics. Whereas nonskeptics emphasized the points of similarity between the two theories, thereby enabling them to adopt insights, intuition, and calculational tools borrowed from the better understood electromagnetic theory of radiation, the skeptics tended to focus on those points at which the analogy broke down, and they did so not merely for rhetorical purposes. An analogy that is perfect in every detail lacks the fertility which permits new insight into the nature of the new phenomenon to be developed. There must be a point where the new, unfamiliar theory becomes distinguishable from the old and takes on its own life and reality. As an example, the quadrupole formula, whose history we will discuss later at great length, describes the rate of energy loss by a source of gravitational waves. It can be derived by a calculation that is analogous to those used in the case of electrodynamics, and the result itself bears comparison with similar results in the case of an electromagnetic transmitter. But there is a striking difference in the gravitational case, which is indicated by the name of the formula itself. For gravity waves, quadrupolar radiation (meaning emission by a system with motions about two separate axes of symmetry, as opposed to one in the case of dipole radiation) is the lowest order of emission, whereas electromagnetic waves exist with dipole symmetry. In fact, our experience of electromagnetic radiation is dominated by dipole sources, and anyone with some experience with radio antennae knows what a dipole is. But gravitational dipoles do not emit waves, which is one of the reasons why gravitational waves are so weak that detecting them has been a mammoth billion-dollar enterprise which is still in progress.

Therefore, a successful analogy must not only exhibit many appropriate parallels between the two objects of comparison, but in order to be fruitful, it must also have points at which the comparison either breaks down or becomes less than straightforward. Given this requirement, one can expect to see different styles of analogic reasoning, in which greater or lesser emphasis is placed on the analogy or its breakdowns. I see the skeptics—among whom I number Infeld, Peter Havas, Arther Stanley Eddington, Nathan Rosen, Hermann Bondi, and others—as those who employed the analogy with electromagnetism in a negative way, while others—such as

John Wheeler, Lev Landau, and Felix Pirani—preferred a more positive use of the analogy. Both approaches are fertile, and the skeptics, who for a time included Einstein himself, made many contributions to the evolution of gravitational wave theory. This is especially true of Bondi, who, in direct contrast to that of a nonskeptic, exhibited his style of analogic reasoning in the following encounter with Wheeler, from the Chapel Hill conference *Conference on the Role of Gravitation in Physics* in 1957.[3] This exchange gives a perfect illustration of the contrast in the positive and negative styles of using analogy. Introducing the session, Bondi remarked:

> The analogy between electromagnetism and gravitational waves has often been made, but doesn't go very far, holding only to the very questionable extent to which the equations are similar. The cardinal feature of electromagnetic radiation is that when radiation is produced the radiator loses an amount of energy which is independent of the location of the absorbers. With gravitational radiation, on the other hand, we still do not know whether a gravitational radiator transmits energy whether there is a near receiver or not. (DeWitt 1957, p. 33)

Clearly, to Bondi the formal analogy between the equations of general relativity and those of electromagnetism (between the Einstein equations and the Maxwell equations), which had first been emphasized by Einstein himself, was not very compelling. He was looking for a more intuitive use of analogy and appeared to find the negative aspects of the comparison more illuminating than the positive ones.

Bondi was interested in the question of whether a person taking exercise by waving two dumbbells around in an unpredictable fashion transmits gravitational waves that carry both energy and information. Note that Bondi doubted that a system of masses moving in a predictable way, such as a man falling or a binary star system (the two elements of which are essentially falling towards each other), could radiate energy. He did carry the analogy with electromagnetism so far as to show that an *induction* of energy in one systems of masses by a nearby one was possible in gravitation. In his presentation at Chapel Hill he discussed whether a cylindrically symmetric system could lose mass as a result of the emission of gravitational waves. He concluded by answering his opening question (above) in such a way as to emphasize differences between the electromagnetic and gravitational fields:

> "To my mind the electromagnetic field is like money spent. I do not get it back unless someone is very charitable. The gravitational field is more like my breathlessness when I do my exercise. When I stop, I regain my breath. If I do not stop (as in the periodic case) I will collapse. In the finite case, which to my mind is the more physical one, no irreversible change has taken place." (DeWitt 1957, p. 36; quotation marks in original)

(For many of the people I interviewed for this book, the image of Bondi doing his exercises to generate gravitational waves was their strongest memory of this conference and others of the period.)

Bondi had presented the case of a cylindrically symmetric system as an example of a system in which the gravitational field does not radiate although it can transmit energy by a form of induction, in analogy with electromagnetic induction. John Wheeler's response to this was to find a deep analogy between the two fields, precisely where Bondi perceived a critical difference:

> "How one can think that a cylindrically symmetric system could radiate is a surprise to me. There seems to be a far-reaching analogy between this case and the problem of emission of electromagnetic radiation from a zero-zero transmission in an atom or nucleus. The charge can oscillate spherically symmetrically, but the system doesn't radiate. However, if we have an electron in the neighborhood, internal conversion can take place, with still no electromagnetic radiation emitted. This could correspond to the uptake of energy of the gravitational disturbance created by the 'cylindrical symmetric' exercise of yours." (DeWitt 1957, p. 36; quotation marks in original)

In response to this, Bondi agreed "he has had suspicions on that side also. To put it crudely, what stops the electromagnetic radiation in the atom is the law of conservation of charge, and what stops gravitational radiation from taking place is the conservation of mass and of momentum. But he does not think there is necessarily anything against radiation of cylindrical symmetry" (DeWitt 1957, p. 36). Note that although the conservation of mass is analogous to conservation of charge, there is no direct electromagnetic analog of conservation of momentum. Bondi reworded Wheeler's example in such a way as to point up the *dissimilarity* between the two cases, rather than Wheeler's "far-reaching analogy."

In the closing remarks for the session, Wheeler elaborated on the analogy between the two fields, nicely summing up the nonskeptic's view of gravitational waves:

> As concerns the radiation problem, we would like to know what is the highest degree of symmetry one can have in a problem, and still have interesting radiation. This leads one to the question of whether, even in the cases where there is no symmetry, one has reason to expect radiation. On this score, it would be well to recall an important physical fact: that the gravitational field of a point charge has close analogies to the electrical field. One knows that there is a certain linear approximation to the field equations similar in nature to the electromagnetic equations, so that if a mass is accelerating, one finds it produces radiation similar to the electromagnetic radiation of an accelerating charge. On this account, one expects gravitational radiation. Using this analogy, Einstein was able to calculate the rate of radiation from a double star. (DeWitt 1957, p. 45)

As we shall see, it is a characteristic of skeptics, from Eddington by way of Bondi to Havas, to deny that Einstein's calculation can be applied to the case of a binary star system.

Wheeler went on to address Bondi's earlier remarks, once again recasting them with a positive spin on the electromagnetic analogy:

> BONDI has reminded us that if one looks for radiation pressure on a particle in gravitational waves, he must take into account the radiation produced by the motion of the particle itself. The situation here is analogous to an electromagnetic wave passing over a particle. To the lowest approximation, the particle only feels the electric field and oscillates with it. If one improves his approximation, he finds that the particle begins to respond to the magnetic field, and moves in a figure eight. Still there is no radiation pressure. It is only when one includes the radiation that the particle itself gives out that one gets radiation pressure. That is, it is only when one allows for the radiation damping force that one finds the particle moving forward. Similarly,

in the case of gravitational radiation, one faces similar problems. As WEBER brought out [referring to an earlier talk by Joseph Weber, the pioneer of gravity wave detection already referred to], in the case of a cylindrical wave, a gravitational metric charge passing over a particle leaves the particle with its initial energy after the disturbance has left. At first sight, one might believe that there is no observable consequence of the action of the wave on the particle. However, the electromagnetic analogy suggests that if one were to go further, one might expect to find radiation pressure. (DeWitt 1957, p. 45)

Wheeler's instinct was to press forward with the analogy, refusing to let go of a useful guide, until all basic problems of understanding gravitational waves had been overcome. Obviously his approach tended to presume their existence, and others preferred to remain agnostic, feeling their way forward, on the lookout for breakdowns in the analogy, fearing that over reliance on it might lead them into pitfalls of understanding.

So in this amusing and revealing interchange, for every similarity between the two fields, electromagnetic and gravitational, seen by Wheeler, Bondi saw a dissimilarity. For every confirmation of the analogy, as Wheeler would have had it, Bondi saw it, on the contrary, breaking down. Which one was right? In fact, Bondi and Wheeler had among the keenest intuitions on the workings of the gravitational field of all the relativists of that time. They and their groups contributed more than any others to the understanding of the phenomenon of gravitational waves that emerged in the years after this interchange. Nevertheless, there was a clash of styles apparent in their approaches to the problem. The question of the analogy between electromagnetism and gravity lay at the heart of the problem of gravitational waves. In general, the skeptics were those who doubted or mistrusted the analogy and used it gingerly and with caution. That they nevertheless used it is, as I said, an important point to keep in mind. Bondi was clearly using the analogy as a guide about as much as Wheeler was, but he was searching for different things. He eagerly looked for the point of breakdown, seeking it out, in the hopes of finding the new and unexpected. Wheeler was hoping to ride the analogic horse as far as it would take him into strange new worlds of physics, beyond experiment but not beyond the imagination. Their methods differed, but their motivation was the same.

2

~~~~~~

The Prehistory
of Gravitational Waves

Gravitational waves are a quintessentially twentieth-century idea, although their actual detection has had to be deferred to the twenty-first century for technical reasons. Their existence was fittingly, but indirectly, proven before the close of the last century; however, a phenomenon now known to be closely related to gravitational radiation was discussed by one eminent theorist as early as the eighteenth century. In this chapter we will discuss the first gropings towards the idea of gravitational waves, which took place before Einstein's general relativity theory of 1916, which properly inaugurated the subject.

As a preamble, we can say that there are two ways of looking at what happens to a source of gravitational waves when it radiates energy. The first is to calculate how much energy the gravitational waves are carrying away and then make use of the principle of conservation of energy to work out the effect the loss of such a quantity of energy would have on the motion of the source. A second method is to make a detailed study of the interaction between the various parts of the source itself. If the time lag caused by the finite speed of propagation of the gravitational force across space is taken into account, then this will cause the motion of the source to damp, or decrease, and it loses energy. Thus, there are two ways of approaching the theory of the source: via a calculation that focuses on the waves as they travel far away from the source system, or via one that deals with the "problem of motion" of the source itself. Thus the problem of what we now call radiation damping, which causes systems like binary stars to decay in their orbital motion as a result of emitting gravitational waves, can actually be attacked without any understanding of the concept of gravity waves at all. All that is required is a study of the problem of motion of such a system, taking into account the time lags (by using what is called retarded time) arising from the time it takes the gravitational force to travel between the two bodies.

During the eighteenth century, the theory of celestial mechanics based on Newtonian gravity was developed to such perfection that in its day it was seen as the pinnacle of the triumph of science and the intellect, just as Einstein's theory of gravity is in our own time. Oddly enough, the most signal triumph of each theory, in which a famous intractable problem of celestial mechanics was overcome by a prodigious intellectual feat, was in each case preceded by attempts to explain the problem in terms of what we would now call gravitational radiation damping. When gravitational radiation (or any form of radiation) is emitted by a system, it is presumed to carry energy with it, and this energy must be balanced (if the law of conservation of energy is to be preserved) by a loss of energy in the emitting system. Whatever motion in the source caused the radiation to be emitted should thus decrease in amplitude or strength. A physicist would say the motion is *damped*. Damping of the motion of a system due to its emission of radiation is referred to as radiation damping, or radiation reaction, or back action. The later two terms may also refer to other radiation-related effects, but in our story there is not much distinction to be drawn between the terms.

The first intractable puzzle, the one that led to a celebrated success for Newton's theory of gravity, was the problem of the Moon. The theory of universal gravitation was first applied to the problem of the lunar orbit by Newton himself, although he was not completely satisfied with his efforts. He later recalled that "his head never ached but with his studies on the moon."[1] The Moon's motion presented a number of calculational difficulties for Newton's gravitational theory. First of all, the Moon's orbit around the Earth is more like a true binary system orbit than any of the planetary orbits. The Moon and the Earth being of comparable size, they can both be said to orbit a common point, rather than the satellite approximately orbiting the central body, as with each of the planets and the Sun.[2] This presents a true two-body problem, which Newton was fully capable of solving. Additional difficulties are presented by the fact that the Sun and the planets act on and perturb this system through their gravitational effects. This many-body problem was not solvable in Newtonian mechanics then or later, but approximate solutions could be obtained through the method of perturbations. In this approach, the effects of the other bodies beyond the second are treated as relatively small, so that the equations can be written as sums of an infinite number of terms in successively higher powers of some small parameter. The terms containing higher powers of this parameter are neglected, so that only the first few terms in the sequence need be evaluated. This technique, which was to be perfected in subsequent centuries, encounters difficulties when some of the perturbations are not exactly small. This is true of the Earth-Moon system, in which the influence of the Sun is an appreciable fraction of the influence of the two main bodies on each other. The Sun is far away but is so large relative to the Moon and the Earth that, for example, its influence on the tides is quite noticeable. It is a well-known phenomenon that the size of tides varies with the position of the Sun, as seen from Earth. What this means is that the approximation which treats the Sun's effect on the motion of the Moon as a small fraction of that produced by the Earth's gravity is not a very good one, and this is one of many issues that greatly complicate efforts to predict the behavior of the Earth-Moon system mathematically.

So the Sun wrestles with the Earth for influence over our satellite, an influence that is continually altered by the motion of the Moon away from and towards the Sun as it orbits us, and by the variations in the Earth's own distance from the Sun. Johannes Kepler was the first to suggest that the Sun exerted an attraction on the Moon that was responsible for some of the variations in its motion. In essence, one is dealing with a three-body problem with additional effects from the other planets, principally Venus. The three-body problem has never been generally solved in mechanics.

Indeed, the problem of the Moon may have helped convince Newton that the solar system could not be stable to many-body perturbations, and that deistic intervention would be required to restore stable initial conditions through regular interventions over the millennia. Among English theologians of Newton's day, there was great resistance to the idea of an eternal universe, which was associated with atheistic thought and the mechanical philosophy of René Descartes. The strong millenialist tradition in seventeenth-century Puritan England was, no doubt, greatly influential in this. Although continental thinkers such as Gottfried Leibniz viewed ideas of a finite "imperfect" cosmos as an insult to its creator, many devout Englishmen feared the idea of a universe in which God would have no reason for existence as an open invitation to atheism. Some English philosophers even resisted the doctrine of inertia, preferring to rely on God to maintain motion in the world (Kubrin 1995). Newton, who was himself fascinated by millenialist ideas, appears to have shared this English prejudice. Indeed, he may have viewed the disorder arising from many-body perturbations as a literal form of dissipation, by which the amount of motion in the solar system would inevitably decrease (Kubrin 1995). He played with various mechanisms through which God would eventually step in, perhaps via some mechanical process, and reform the cosmos.

The first astronomer to uncover actual evidence of long-term alteration in the celestial motions was Edmond Halley, who examined records of medieval solar eclipses made by the Arab astronomer al-Battani (known to the Latins as Albategnius), as well as ancient eclipses reported by Ptolemy, and discovered apparent discrepancies of the order of an hour in the eclipse times by calculating backwards from contemporary lunar positions on the basis of the known lunar period. Halley speculated that, if the accuracy of al-Battani's latitude estimations could be confirmed, there was evidence for a secular longitudinal acceleration of the moon, meaning that the moon must, in the earlier epoch, have been moving longitudinally (i.e., across the sky from east to west) more slowly than it was in Halley's own time. If this effect existed, it would be of considerable interest, since it would show secular, and not periodic, change in the motion of one of the principal celestial bodies. A secular change, unlike a periodic one, always operates in the same direction, so that the motion never returns to its starting point. Thus secular changes in orbital systems can ultimately result in the demise of the system itself. Halley himself speculated that the change might be due to an increase in the mass of the earth, the consequence of Newton's idea that the earth attracted the ether of space (visible in the form of cometary tails) into itself by the force of gravity and thus continually augmented its mass.[3]

That such a speculation should first be made in England is not surprising. Indeed, it seems that when he was a candidate for the Savilian Chair of Astronomy at Oxford in 1691, Halley was forced to defend himself against the charge of upholding the doctrine of the eternal mechanical cosmos (Armitage, 1966). Although he was not awarded the post, the charges against him appear to have been unfounded (it was also alleged that, like Newton, he was a Unitarian). Certainly he was the only scientist or philosopher of the time able to advance, by his historical analysis, evidence for the decrease of motion in the solar system, and therefore of decay in the cosmos. Given how little evidence he had to go on, we may wonder whether he was rather predisposed to a conclusion that showed evidence for signs of cosmic decay. However sharp the difference in outlook between England and the continent may have been, Halley's secular acceleration of the Moon eventually did become an accepted fact in eighteenth-century astronomy, after further contributions from the English astronomer Richard Dunthorne, the German Tobias Mayer, and the Frenchman Joseph-Jérôme Lalande (Armitage 1966). This intriguing effect, discovered in the historical archives, became one of the best-known puzzles in celestial mechanics up to the present time.

Prizes played an important role in the economy of eighteenth-century science, especially in the development of the theory of the Moon. The most lucrative prize was the British government's offer of up to £20,000 for a method of accurately determining longitude at sea. A number of methods that had been suggested over the years involved celestial observations from on board ship, and since the use of a telescope was impractical from the heaving deck of a ship underway, it followed that the Moon, readily visible with the naked eye, was the best independent celestial clock available. However, even Newton's lunar theory was inadequate to predict the erratic motions of the moon to the accuracy required for navigational reference. Either better observations of the complete lunar cycle or an improved lunar theory would be needed, preferably both.

Within the academic sphere, the Paris Academy of Sciences offered prizes for solutions of problems outstanding in Newtonian gravitational theory several times in the 1760s and 1770s. One of these problems, involving Newton's formula for the motion of the perigees, had already led Leonhard Euler to suggest that a modification of the basic Newtonian theory might be necessary to save the phenomena. This proved unnecessary after Alexis-Claude Clairaut and Jean Le Rond d'Alembert, following years of acrimonious dispute between them over the prize, each produced a solution (Peterson 1993). Euler was obliged to have his own St. Petersburg Academy of Sciences offer another prize (tempting Clairaut to resubmit his solution) before he could discover how the Frenchman had accomplished it.

Euler himself received a modest partial share of the longitude prize for his theoretical work, together with Tobias Meyer, for his observational work.

The Paris academy's prize of 1773 sought an explanation of the secular acceleration of the moon, whether as a result of perturbations produced by the Sun or planets, or by the non-sphericity of the Earth. Once again the problem proved exceedingly difficult, and Lagrange won it for a brilliant thesis on perturbation theory. His conclusion was that the effect could not be explained by perturbations within the Newtonian theory. Several alternative hypotheses were offered. Euler thought that a subtle medium in space retarding the Moon's motion might be responsible, while Kant suggested that a tidal friction of the Moon acting on the Earth might explain the acceleration, since as the day lengthened, all celestial motions would appear quicker as observed from Earth (Felber 1974; Brosche 1977). However, since there was no evidence for secular acceleration in the Sun or other planetary bodies, this suggestion was not generally adopted.

Pierre-Simon de Laplace, still young but destined to become the greatest exponent of celestial mechanics, took a systematic approach to the suggestion that an alteration in the basic theory of gravity might be required. In a paper of 1776 he suggested four fundamental ways in which the theory might be modified: the inverse-square law, universality, instantaneous propagation, and the equivalence of attraction for bodies at rest and in motion. Pursuing the last two suggestions (which are clearly linked), he calculated the effect on a simple orbit of assuming a finite propagation speed of gravity. His conclusion was that it would result in a decrease of the orbital radius, and a resultant accelerated longitudinal motion, but that the effect, if entirely due to this cause in the case of the moon, would indicate a speed of gravity 7 million times that of light. This formidably high speed, still hardly distinguishable from instantaneity, was no compelling reason to take up the idea of finite propagation.

Laplace's calculation conceives of the gravitational force as mediated by a corpuscle passing between the attracting masses (see figure 2.2). If the Moon is orbiting the Earth and emits such a corpuscle, it must aim a little ahead of the Earth's present position in order to strike the latter body, if the corpuscle travels with a finite speed. This means it must be emitted not only in a "downward" sense, but in a slightly "backwards" direction (relative to the lunar motion). Since the direction of emission of this corpuscle indicates the direction of the gravitational pull exerted on the

Figure 2.1. Pierre-Simon de Laplace, the first scientist to explore
the idea that a finite speed of propagation of gravity would result
in damping of otherwise stable orbits. (Courtesy AIP Emilio
Segré Visual Archives)

Moon by the Earth, the Moon will not only be attracted towards the Earth,
but also be impeded somewhat in its tangential motion by the emission
of such noninstantaneous particles. This loss of angular momentum will
force it to move inward in its orbit (falling towards the Earth), which in
turn increases the rate of its longitudinal motion, as seen from Earth. The
retarding effect clearly depends on the angle that the direction of emission
makes with that of the instantaneous central force. This is simply v/c,
where v is the lunar velocity, and c is the velocity of the corpuscle.[4] This is
perhaps the first radiation reaction calculation in the problem of motion.
But of course, Laplace's corpuscular view of gravitation had no place for

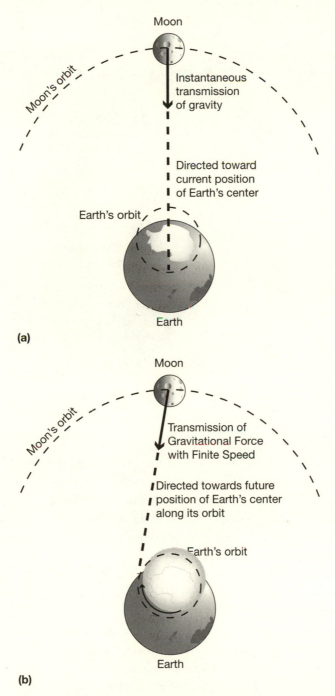

Moon

Instantaneous
transmission
of gravity

Directed toward
current position
of Earth's center

Moon's orbit

Earth's orbit

Earth

(a)

Moon

Transmission of
Gravitational Force
with Finite Speed

Directed towards future
position of Earth's center
along its orbit

Moon's orbit

Earth's orbit

Earth

(b)

Figure 2.2. Laplace's orbital damping. (Figure continued on
p. 26.)

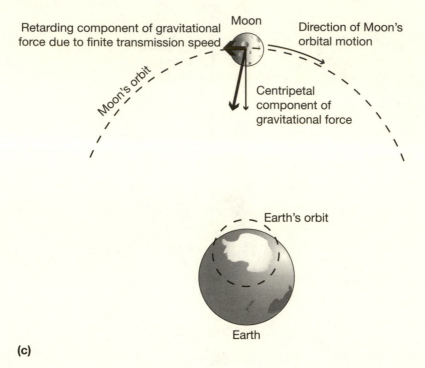

(c)

Figure 2.2. (Continued)

the radiation emission side of what has been called the Laplace effect, orbital decay caused by a noninstantaneous force of attraction. One must keep in mind that the law of conservation of energy did not then play the important role in physics that it came to in the nineteenth century. It did not necessarily occur to Laplace to ask where the kinetic energy lost by the Earth-Moon system was disappearing to.

It seems likely that the problem of the secular acceleration of the Moon continued to attract attention not only because it seemed a possible breakdown of Newtonian theory, but also because it seemed to be a definite example of decay in the heavens. We have already seen how in the England of Halley's day, the doctrine of an eternal creation was viewed in some quarters as a grave heresy. This millenialist style of thought does not seem to have been so prevalent on the continent, and we can speculate that proposals such as those of Euler, Kant, and Laplace, which sought to explain the acceleration of the Moon by an appeal to dissipative effects, may not have found favor because of an uncomfortable feeling that the world system ought to be eternal. Furthermore, the analogy with the human clockmaker, which in Newton's day depicted the deity as an artificer obliged periodically to rewind the mainspring of his creation (or reset the pendulum) and set it once more in ordered motion, while appropriate to the rather unreliable clocks of Newton's day, seemed an even less flattering portrait by the end of the eighteenth century (Leibniz had already criticized that view as a slur on the Almighty in Newton's time). The problem of longitude, which had inspired such detailed study of the Moon, in theory and in experiment, had also driven a comparable improvement in the science of chronometry. Indeed, a method of finding longitude at sea that did not require clear skies, complex charts and tables, and astronomical instruments, was provided by an English watchmaker, John Harrison, who invented a series of clocks, followed by a portable watch, that could tell time with marvelous accuracy over months at sea and in the shipboard environment (Sobel 1995). By Laplace's time, it no longer seemed appropriate to conceive of a creator so unskilled as a craftsman that he would construct his world system only to see it fall into ruin and disorder over the course of its own action.

Subsequently, however, Laplace discovered a complex perturbative effect that had been missed by Lagrange and that not only gave a nondissipative explanation for the entire acceleration of the moon, but in fact showed the motion to be periodic, although with a period of millions of years. Lagrange, who considered that no combination of the other planets and the Sun acting on the lunar orbit could explain the acceleration, had missed the perturbation because the other planets acted only indirectly on the Moon. In Laplace's explanation, the net effect of the planets was on the *Earth's* solar orbit so as to reduce its eccentricity by small amounts over centuries. This change altered, in the mean, the Moon's position with respect to the Sun over its orbit, reducing the net amount by which the Sun tends to draw the Moon away from the Earth. Thus the Moon gradually approaches the Earth by just such an amount, as Laplace calculated it, as to precisely account for the observed decrease in its orbital period. Eventually, however, the complex effect would reverse itself and begin once more to draw the Moon away from the Earth. Therefore, in *Celestial Mechanics*, Laplace was able to present his solution as a tour de force, capping his vindication both of Newtonian theory and of the eternal clockwork universe concept, by showing the stability of the system of planetary orbits to its own perturbations. When his explanation of the secular acceleration was taken into account, his "back reaction" calculation was now only presented as proving that the action-at-a-distance assumption was justified. In view of the absence of any acceleration of the Moon in excess of the prediction of his perturbation theory, Laplace concluded a minimum speed for the propagation of gravity of 100 million times that of light (Laplace 1825). This result was very well known in the nineteenth century, as is evident from Poincaré's paper discussed below and as is made clear in a recent study of gravitation theories in the eighteenth and nineteenth centuries: "During the nineteenth century these calculations were often presented as an (almost) insurmountable obstacle to all explanations of gravitation based upon the action of an intermediate fluid" (Van Lunteren 1991).

Despite its onetime fame in scientific circles, Laplace's explanation of the acceleration of the Moon has not survived unaltered down to our own day. In the mid-nineteenth century, the English astronomer John Couch Adams recalculated Laplace's effect and showed that, owing to the neglect of certain terms which, in fact, added up to an appreciable sum, Laplace's effect was only half what Laplace himself had calculated it to be. In destroying the perfect agreement with observation of this famous result, Adams precipitated a fierce controversy with Urbain Leverrier, among others, whose passions were further inflamed by the nationalistic rivalry between English and French science then prevalent. However, his correction of Laplace did lead to the revival by Charles Delaunay, Leverrier's leading French rival, of Kant's tidal friction idea. His calculations showed that the slowing of the Earth's rotation by this force could account for the remaining half of the effect. It was not, however, until the twentieth century that a corresponding acceleration of the Sun was observed.

Today it is well known, from laser range-finding made possible by a mirror placed on the Moon by an Apollo mission, that the Moon is in fact receding from the Earth, not approaching it. The explanation for this is found in the phenomenon of tidal friction, which, despite its long pedigree, first became generally accepted only thirty to fourty years ago.[5] The Moon, as is well known, raises tides upon the Earth's oceans, both directly below it and at the antipode of that point. The Earth's rotation, however, drags the tidal bulges somewhat ahead of these idealized positions, typically by about 3 degrees (see figure 2.3). This is known as tidal lag, since it means that an observer on the land will see the Moon overhead before she experiences the high tide. The near bulge naturally exerts its own gravitational attraction on the Moon and tends to pull the Moon somewhat forward of its own position. The bulge on the other side of the planet has a retarding impulse, of course, but its effect is smaller, as it is farther away. The net effect of the tidal bulge is to impart an increased forward momentum to the Moon. This increase in angular momentum forces the Moon upwards in its orbit and is gained at the expense of the Earth's rotational angular momentum, which is slowed by the tidal bulges' attraction towards the Moon. The recession of the Moon from the Earth would cause us to observe a longitudinal *deceleration*, except that our own clock is slowing down with the centuries at such a rate that we conclude that the Moon is moving faster than hitherto. In other words, the month has lengthened, but the number of days it contains has decreased, due to the lengthening of the day. This is why Halley thought that the month had grown shorter with time.

It is hardly surprising that Laplace did not have the last word on so complex a subject as the theory of the Moon's orbit, and the same holds true of celestial mechanics and perturbation theory in general. Throughout the nineteenth century there were continued refinements in the theory and observation of planetary motions. Indeed, it is during this period that the most famous achievement of Newtonian gravitational theory occurred, the prediction (by Leverrier and independently by Adams) and discovery (by Johann Galle) of the planet Neptune. By the end of that century, the most prominent anomaly in celestial mechanics was no longer associated with the nearest body to the Earth but rather with the nearest planet to the Sun, Mercury. According to standard Newtonian theory, the unperturbed orbit of a planet should have it return to its closest approach to the Sun at the same angular position in each orbit, but instead, Mercury was observed to shift its perihelion around in its orbit by 43 arc-seconds per century, in excess of what could be explained by perturbations from the other planets.

Various explanations within Newtonian theory that would explain the effect had been suggested over the years (for the history of this fascinating topic see Roseveare 1982; Earman and Janssen 1993; and Janssen 2006), the most famous being the hypothetical planet Vulcan, nearer than Mercury to the Sun, whose perturbative effect on the latter would account for

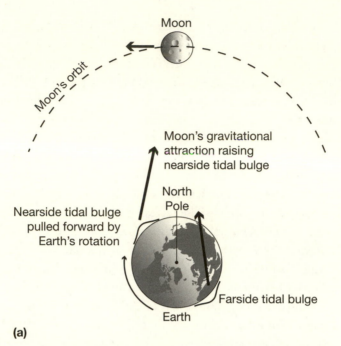

(a)

Figure 2.3. Tidal friction. (a) The Moon's gravitational pull raises tides on the parts of the Earth nearest to and farthest away from it. But the Earth's rotation drags the bulges forward producing what is known as "tidal lag," because we see the Moon overhead before we experience the high tide it causes. (b) *Top*: The two tidal bulges produce a net gravitational tug on the Moon, which pulls it down and forward, accelerating it. The result is to increase its angular momentum, which comes at the expense of the Earth's rotational angular momentum. As a result, the Moon draws further away from us. *Bottom*: The Moon's pull on the tidal bulges produces a torque on the Earth that opposes its rotation, slowing it down and lengthening the day. Thus, all motions in the heavens appear more rapid to us. (c) As the Moon spirals outward from the Earth, the month grows longer as the Moon's apparent motion across the sky is slower. But at the same time, the Earth's rotation slows, increasing the length of the day so that there are fewer days in a month. Thus, it appears as if the month grows shorter.

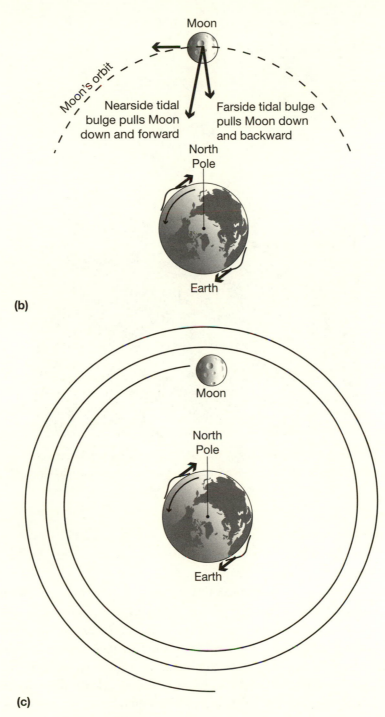

(b)

(c)

Figure 2.3. (Continued)

Figure 2.4. Henri Poincaré was, with Laplace, among the greatest theorists of celestial mechanics. He was also a pioneer in discussing the possible existence of gravitational waves. (Courtesy AIP Emilio Segré Visual Archives)

the shift. Vulcan was searched for repeatedly, but unsuccessfully, by several astronomers towards the end of the ninteenth century. One alternative approach, as with the eighteenth-century problem of the Moon's secular acceleration, was to posit changes to the theory of gravity. In 1908 Henri Poincaré, the greatest theorist of celestial mechanics and perturbations since Laplace, made a radical (but tentative) proposal: that the emission of gravitational waves from the orbit of this quickly moving inner planet was removing sufficient energy from its motion as to show up in the form of the perihelion shift. He based the idea on an earlier calculation of another noted French astronomer, François-Félix Tisserand.[6]

That Poincaré should introduce the idea of gravitational waves at this time was due to the influence of Maxwell's theory of electromagnetism, which successfully predicted the existence of electromagnetic radiation in the second half of the nineteenth century. Poincaré used the term *gravitational wave* as early as 1905, and the term was probably used by others even earlier (see Katzir 2005, for a discussion; the earliest usage he mentions is from 1902). However, it appears that these early uses typically referred only to the fact that, in a field theory, the gravitational effect must propagate with a finite speed, contrary to Laplace (Katzir 2005, p. 22).

By 1908 the efforts to reformulate Galilean relativity to accommodate Maxwellian electrodynamics had led to a deepening appreciation of the role played by radiation in field theories. Since the speed of light was a key parameter in the new relativistic equations for electrodynamics, it seemed reasonable to speculate by analogy that if this new form of relativity were to apply to gravity, there must exist some form of "gravitational radiation," which would propagate at the speed of light. Just as in the electrical case, where an accelerating charge would emit radiation and brake its own motion in consequence, so a massive body orbiting the Sun could be expected to lose energy to some as yet unknown type of radiation (Poincaré referred to it as the *onde d'acceleration* or "wave of acceleration"). Such an effect might account for hitherto anomalous effects such as the perihelion shift of Mercury. As the nearest planet to the Sun, and thus the fastest moving planet, Mercury could be expected to lose more energy by this mechanism than any other and thus exhibit most strongly any associated effect on its orbit.

Now, in discussing (in *Science and Method and The New Mechanics*) the question of how and whether the principle of relativity (meaning more or less what we would now call special relativity) should be applied to gravitational forces, Poincaré was well aware of Laplace's result that the propagation of gravity must take place at a speed 100 million times that of light. However, he preferred to regard Laplace's result as largely unsubstantiated and instead proceeded to a discussion of the implications of modern relativity for gravitation:

> Are the foregoing theories [Lorentz's unification of the gravitational and electrostatic forces] reconcilable with astronomical observations? To begin with, if we adopt them, the energy of the planetary motions will be constantly dissipated by the *wave of acceleration*. It would follow from this that there would be a constant acceleration of the mean motions of the planets, as if these planets were moving in a

resisting medium. But this effect is exceedingly slight, much too slight to be disclosed by the most minute observations. The acceleration of the celestial bodies is relatively small, so that the effects of the wave of acceleration are negligible, and the motion may be regarded as *quasi-stationary*. It is true that the effects of the wave of acceleration are constantly accumulating, but this accumulation itself is so slow that it would certainly require thousands of years of observation before it became perceptible. . . .

It is in the motion of Mercury that the effect will be most perceptible, because it is the planet that has the highest velocity. Tisserand formerly made a similar calculation . . . and found that if Newtonian attraction took place in conformity with Weber's law [a nineteenth century nonlinear theory of electrodynamics], there should result, in the perihelion of Mercury, a secular variation of 14″, *in the same direction as that which has been observed and not explained*, but smaller, since the latter is 38″. (Poincaré 1908, pp. 239–240)

Poincaré's own calculation, based on the relativistic theories of Lorentz and Max Abraham, predicted a smaller effect, again in the same sense as the observed effect, of 7″ per century and 5.6″ per century respectively for Mercury. It seems clear that his use of the term *onde d'acceleration* means, in this context, what we would now call gravitational waves. This identification is made a little problematic by Poincaré's use of the same term to describe electromagnetic emission from accelerating charges and by his employment of a unified theory of gravitation and electromagnetism, but, as the context is the extension of the relativistic principle to gravitation and accelerated motion, we can be quite confident of his meaning:

To sum up, *the only appreciable effect upon astronomical observations* [of extending the principle of relativity to gravitation] *would be a motion of Mercury's perihelion, in the same direction as that which has been observed without being explained, but considerably smaller.*

This cannot be regarded as an argument in favor of the new Dynamics, . . . but still less can it be regarded as an argument against it. (Poincaré 1908, p. 242)

(For more on Poincaré's ideas on a relativistic theory of gravity, see Katzir 2005.)

In the end, of course, the perihelion shift (like the lunar secular acceleration before it) proved to be a conservative effect. Its explanation by Einstein's new general theory of relativity was the most striking initial achievement of the theory and helped make it the most famous scientific achievement of the century and Einstein its most famous scientist. Unlike Laplace's explanation of the Moon's secular acceleration, Einstein's remarkable result has not since been overturned. In 1938 Einstein and his collaborators (see chapter 8 for a discussion of the Einstein, Infeld, and Hoffmann paper) produced a post-Newtonian theory of orbital motion based on general relativity, while another colleague, Howard Percy Robertson, employed the new scheme to recalculate Mercury's perihelion shift, again agreeing with the observed excess from perturbations (Robertson 1938). Attempts were made in the 1950s by Dicke and others to explain part of the effect as the result of a large quadrupole distortion of the Sun, which would have cancelled the agreement with general relativity and instead, perhaps, vindicated the rival Brans-Dicke theory (Brans and Dicke 1961; see also Kaiser 2006 for aspects of the history of this theory), but to date, such efforts have not been successful (Will 1993).

The most prescient remarks on the subject of gravitational waves before 1916 are to be found in the work of Max Abraham. Abraham was a near contemporary of Einstein's, like him a brilliant Jewish theoretical physicist who encountered difficulty in getting a job in the anti-Semitic world of German academia. The early part of Abraham's career was much more successful than Einstein's. He was able to continue as a professional physicist after graduation, though with much difficulty. Abraham was a dedicated proponent of the electromagnetic worldview, a unification program that sought to reduce all of physics to phenomena arising out of the electromagnetic field and the ether. He saw relativity theory, as promulgated by Einstein, as a serious intellectual and professional threat. The two men were rivals from the start and remained so.[7] After the publication of Einstein's early work on a relativistic theory of gravitation, Abraham took up similar work and produced a theory of his own in 1912. His theory was vigorously criticized by Einstein and faded completely from memory within a short time. But in his paper he discussed the existence of gravitational waves and concluded that they could play no significant role in a relativistic theory of gravitation. He was thus the first skeptic.

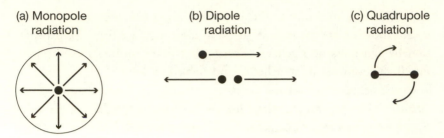

(a) Monopole
radiation

(b) Dipole
radiation

(c) Quadrupole
radiation

Figure 2.5. Multipole radiation. (a) Monopole radiation can only be caused by a change in the source that is perfectly spherically symmetric. But purely radial oscillation as shown does not change inverse-square law fields such as electromagnetism and gravity. A source simply disappearing would produce monopole radiation, but the conservation laws of charge and mass forbid this. (b) An accelerating source, for instance one that moves backward and forward along a straight line, is a dipole source of radiation. The change in the source takes place along one axis only (an axis having two poles). This works fine in electromagnetism. But in the case of gravity, conservation of momentum demands that, for each such moving mass, there is a mass with equal momentum moving oppositely, canceling out any change in the field. (c) Quadrupolar radiation is produced when the source changes about two axes (hence four poles). A classic example of a source of quadrupolar radiation would be a spinning dumbbell shape. Take away the strut, and you have a binary star.

His arguments were entirely sound, and indeed, this is one case where Einstein could have benefited from heeding him, as we shall see. Abraham was a brilliant theorist even though it was his tragedy to line up on the wrong side of the main issues of his day and his field. He was, however, the first person to recognize that there can be no such thing as dipole gravitational radiation. (This discussion is based on the historical research of Jürgen Renn [2006]).

Since waves are traveling disturbances in a medium, it follows that they must find their origin, or their source, in some kind of disturbance or disturbing mechanism. One can classify waves according to the symmetry properties of the source disturbance (see figure 2.5). Suppose it is perfectly symmetrical in every way, then we say that the source has only one "pole" or point of origin. Radiation from such a source is called monopole radiation (the source itself is called a monopole). A disturbance of this sort would be an in-and-out motion of a sphere, for instance. It looks the same from all directions. Now imagine a back-and-forth motion so that the disturbance moves along a line. The situation is no longer *spherically symmetric* as it was in the monopole case, but there is still an *axial symmetry*. If one moves around the source as if pivoting about its axis, everything still looks the same. This source has two "poles," one at either end of the axis, and is known as a dipole. The radiation it emits is called dipole radiation. The characteristic radiation in electromagnetism is dipole radiation; most radio transmitters are primarily dipoles. One can add another axis of disturbance to create quadrupole radiation. A good example of this kind of vibration would be that of a water drop, which wobbles so that first it is tall and thin and then becomes short and fat. And indeed there are octopole and other higher orders of radiation also. By *higher order* we refer not to the fact that each higher pole is at a higher frequency but to the fact that in typical cases where a single source emits radiation at a variety of multipoles, it will emit the greatest part of the energy radiated in the lowest order multipole and successively less and less at each higher order. Thus electromagnetic waves are typically dominated by dipole radiation. But keep in mind that there are also many practical scenarios in which a source may radiate more strongly in higher multipoles than in the lower ones, though these cases, which often involve sources that move at relativistic speeds (close to the speed of light), were not so well known in 1912.

So as one adds more poles, one typically finds that the strength of the radiation decreases. An electromagnetic source that transmits octopole radiation usually also transmits quadrupole and dipole radiation, each at greater strength. What about monopole radiation? Well, both the gravitational and electromagnetic fields are unaffected by the kind of oscillation that sees a sphere shrink and expand. The distant field is the same regardless of the size of the sphere (this result holds generally for fields or forces which exhibit an inverse-square law decline in the strength of their force, and it was first derived by Newton himself). The only way for the field to change is for the sphere of mass or charge to disappear or somehow change its quantity. But the laws of conservation of mass and charge tell us that this never happens (as does our experience of objects in real life). So electromagnetic radiation is dominated by the *lowest order* type of radiation permitted, the dipole.

What Abraham realized is that there is a critical difference between the two fields, electromagnetic and gravitational, and his argument is actually a very Einsteinian one. The principle of equivalence, the cornerstone of Einstein's theory but recognized as true since the days of Galileo, says that inertial mass, which governs how bodies accelerate, and gravitational mass, which governs their gravitational attraction, are always exactly the same. This means (in the context of Newtonian gravity) that things fall towards Earth with the same rate of acceleration no matter how heavy they are or what they are made of. Let us suppose you wanted to create a gravitational wave. To do so you have to make a change to a gravitational field, because when we talk about a wave we really refer to the leading edge of propagation of some change in the field or medium (just as the first news at the edge of a pond that a stone has been thrown into it is when the ripples strike the bank). What is the simplest change we can make to a gravitational field, say the field of the Earth? One way would be to simply make half the Earth disappear. The reduced gravitational mass would be a major change in the planet's gravitational field, and it would have enormous effects on, for instance, the orbit of our Moon. Of course the Moon would only know about the change when the wave arrived at the Moon, but such a wave will, as far as we know, never be detected, since we know it is impossible to make mass simply disappear, because of the law of conservation of mass (even if the Earth were blown up and converted to pure energy that energy would have mass). This kind of a wave we would call a monopole wave, and it does not exist in electromagnetism either, since conservation of charge says that you cannot make electric charges disappear.

If one cannot make charges or masses disappear, one can at least move them, and doing so ought to lead to a change in the fields they produce. However, it turns out that it is the starting and stopping part of the motion that creates the waves. Motion with constant speed (inertial motion) is not generally associated with either electromagnetic or gravitational waves, but accelerated motion is. The amount of charge moving up and down an antenna (the classic dipole in electromagnetism) obviously affects how much electromagnetic radiation is produced, and since the speed at which the charges are moved up and down apparently matters also, it is not hard to be persuaded that the critical quantity we need to change in order to make electromagnetic waves with such a dipole is the product of the electric charge times its speed. Now we can imagine a dipole with two *masses* that we cause to oscillate back and forth. In this gravitational case, the "charge" that generates the field is the gravitational mass which, by the equivalence principle, is just the same as their inertial mass, so that the product of "charge" and velocity of a body is simply its momentum. But the law of conservation of momentum tells us this cannot change. So conservation of mass, for the monopole case, and conservation of momentum, for the dipole case, require that the lowest order of gravitational radiation must be quadrupole.[8] Abraham went only as far as deciding that there is no dipole radiation, and from that he concluded that gravitational waves do not exist, since after all it is dipole radiation which one normally encounters in the case of electromagnetism. Certainly his was the first recognition that gravitational waves were going to be hard to detect, but as we shall see, he was eventually proved wrong in his estimation that they would play no role at all in the theory.

Intriguingly, it seems that Einstein's first reaction on the completion of his theory was to conclude that gravitational waves do not exist. We know this from a letter he wrote to his colleague Karl Schwarzschild on February 19, 1916. Schwarzschild, a leading German astronomer with a strong mathematical background who would shortly after become a casualty of the First World War had just discovered the exact solution to the Einstein equation that bears his name. As he and Einstein discussed this and other matters, such as a public controversy surrounding an astronomer colleague of Einstein, Einstein let slip the following side remark:

> Since then [November 4], I have handled Newton's case differently according to the final theory.—Thus there are no gravitational waves analogous to light waves. This probably is also related to the one-sidedness of the sign of scalar T, incidentally. (Nonexistence of the "dipole".) (Einstein 1998, Doc. 194)

We see that Einstein, like Abraham, realized that there was no such thing as a "gravitational dipole." He deduced this from the well-known fact that there are no negative masses in nature in the way that both negative and positive electric charges exist. There is only one pole to the gravitational force, attraction between masses; there is no such thing as repulsion, or antigravity. The "one-sidedness of the sign of scalar T" refers to the fact that there are no negative masses in the universe. Einstein's equations relate the Einstein tensor, which represents the curvature of spacetime, to the stress-energy tensor, usually written $T_{\mu\nu}$, whose "trace,"[9] written T, represents the stress-energy of a body (in general relativity, the energy of a body, including thermal, kinetic, and other forms of energy; its mass; and mechanical and other stresses inside the body all contribute to its gravitational mass). In what way is this comment of Einstein's related to Abraham's remarks about the nonexistence of gravitational dipole radiation? Well, Abraham's proof depended on the principle of equivalence, which says that all masses fall the same in a gravitational field, but of course, if there were such a thing as negative masses that were repelled by ordinary matter, the first casualty would be the principle of equivalence, since negative matter on the surface of the Earth would fall upwards. It would probably be possible, if negative masses were available, to construct radiating gravitational dipoles out of one negative and one positive mass. It would seem, on the evidence of Einstein's first paper on gravitational waves (which was written a few months after the date of this letter), that Einstein did not develop the germ of this argument in the way that Abraham had.

Einstein's comment to Schwarzschild is unclear about the details of his "discovery" that gravitational waves do not exist. It is apparent that it was not simply based on a simplistic dipole argument, which here he merely adduced as a possible underlying reason. It maybe that we can find a clue in the remark "Newton's case." Some time during the period from November 1915 to February 1916, Einstein had engaged in some calculation which suggested to him that gravitational waves might not exist. The nature of that calculation will be the subject of the next chapter.

3

The Origins
of Gravitational Waves

From the first publication of the Einstein field equations (Einstein 1915), it was obvious that exact solutions of the equations would not be easily found, and it was natural that useful approximation schemes should be sought as an aid to calculation within the new theory. The Einstein equations consist of ten nonlinear equations, for which a fairly large number of solutions have by now been found. However, many of the solutions refer to physically simple or uninteresting situations, and their real utility is often as a basis for providing an approximation scheme to describe more realistic scenarios. Even in the absence of any solution at all, approximation schemes provide a means in which useful calculations can still be accomplished. It may sound odd that one can approximate to a solution that is actually unknown, but here the existence of successful theories which predate general relativity plays a very important role. If we have a theory that we know to be wrong in principle, but which nevertheless provides an accurate description of experiments and observations in practice, then approximating our calculations based on general relativity to the known solutions of the older theory seems a reasonable way to proceed.

Let us take as an example the very first and most famous calculation in the history of general relativity. Since first commencing work on developing a relativistic theory of gravity, Einstein had nurtured the hope that such a theory would explain one of the best-known anomalies of celestial mechanics, the perihelion advance of the planet Mercury. Newton's theory of gravity demands that planets orbiting the Sun move in closed ellipses (a law originally discovered by Johannes Kepler), where by *closed*, we mean that the planet repeatedly moves over the same path in space with each orbit, and the ellipse therefore maintains a constant orientation in space (the long axis of the elliptical shape always points in the same direction). This two-body problem of one planet and the Sun can be solved exactly in the old Newtonian theory. To take account of the gravitational interactions of all the planets with each other requires *perturbation theory*, in which these many small pulls are treated as small adjustments to the large single pull of the Sun on the planet in question. These perturbations to the planet's ideal elliptical motion have the effect of causing the ellipse to precess. One can say that the planet's orbit is no longer closed; instead, it comes back to a different point at each closest approach to the Sun (known as perihelion) and fails to trace out exactly the same path with each orbit. Its path still appears close to elliptical, but the axis of the ellipse is continually changing its orientation. The angle that the axis moves through each year or century determines the rate of precession.

In general relativity, one cannot solve the two-body problem exactly, even today. When Einstein tackled this problem in late 1915, he did not even know of an exact solution for the gravitational field of one body (such as the Sun) by itself. Nevertheless, he proceeded by expressing all of his quantities as series expansions in terms of the speed of the planet Mercury divided by the speed of light (v/c) and other terms depending on the strength of the gravitational field of the Sun. In relativistic terms, which is to say relative to the speed of light, Mercury moves very slowly, so that v/c is a very small quantity, much smaller than one. By a similar token, the gravitational field of the Sun, even as close to the Sun as the position of Mercury, is, in relativistic terms, very small. Now if v/c is very small, it follows that $(v/c)^2$ is even smaller, and so on with higher powers of this quantity. Einstein's 1915 calculation of the perihelion advance of Mercury was developed by expressing everything in terms of, first, terms equal to the Newtonian quantities, and second, in terms of general relativistic corrections containing $(v/c)^2$ (there were no "first-order corrections," of order v/c). All terms of higher order than $(v/c)^2$ (or first-order in the gravitational field of the Sun) were ignored. Although there is no way to be certain that some one or more of these neglected terms might not actually be significant, the smaller the expansion term, the less likely this is, since all higher-order terms are being multiplied by a very small quantity (but recall how Laplace was fooled in his calculations on the longitudinal acceleration of the Moon). These kinds of approximations are generally known as post-Newtonian because they consist of a series of small corrections to results that are identical to the old Newtonian theory. Corrections of order $(v/c)^2$ are known as first-post-Newtonian-order terms (because of the absence of corrections of order v/c). Corrections of order $(v/c)^4$ are second-post-Newtonian. As we shall see, corrections due to the existence of gravitational waves do not enter into such calculations until $(v/c)^5$ ($2\frac{1}{2}$-post Newtonian) order.

Other physicists who read Einstein's first papers on general relativity in late 1915 realized immediately that exact solutions were possible, and the first one to be published was Karl Schwarzschild's solution for the gravitational field of a single body (a solution for what we now refer to as a black hole, but was then simply presumed to be an idealized representation of a star or some other material body). This provided a better way to approximate the orbital motion of a planet such as Mercury, which is so much less massive than the Sun that their combined gravitational field can be treated as simply due to the mass of the Sun plus some small corrections (or perturbations) due to Mercury's mass. So when Einstein demonstrated his calculation of the perihelion advance of Mercury for his course on general relativity given in Zurich in 1919 (Einstein 2002), he naturally made use of this perturbation method based on the Schwarzschild solution rather than on his relatively primitive initial approximation scheme. Much later, in the 1930s, as we will see, he returned to this problem of motion in a much more general way. It often happens that calculations are superseded by more sophisticated work as time goes by, but physicists may still have reasons for believing that the earlier, relatively crude, calculations continue to hold true. For instance, one reason Einstein believed his calculation of the anomalous Mercury perihelion advance was that the result had already been observationally determined. It would have seemed an extraordinary coincidence if it turned out that the astronomical result was wrong, that Einstein's calculation was also wrong, and yet both happened to reproduce exactly the right result.

It was clearly very important that the new theory inherit the theoretical experience and insight gained in years of work on existing theories, such as Newtonian gravitational theory and special relativity. Approximation schemes based on comparisons between general relativity and both Newtonian gravity (Newtonian or post-Newtonian approximations) and special relativity (or relativistic electrodynamics, via the so-called linearized approximation) were important not just to permit calculations but also to show that the theory was a plausible theory of gravity which agreed with the many known measurements and observation on the motions of the planets and other bodies of the solar system. The two main approximation schemes were introduced within a year of the publication of the field equations. These two systems would play important roles in research in general relativity for the rest of the century, especially where astrophysical applications of the theory were concerned. It was the 1916 paper, entitled *Approximate Integration of the Field Equations of Gravitation*, in which Einstein introduced gravitational waves to the theory and the linearized approximation scheme was first developed.

Before discussing this paper, it will be useful to return to Einstein's remark to Schwarzschild from early 1916, and I quote it at greater length:

> My comment . . . in the paper of November 4 no longer applies according to the new determination of $\sqrt{-g} = 1$, as I was already aware. The choice of coordinate system according to the condition $\sum \frac{\partial g^{\mu\nu}}{\partial x_\nu}$ is no longer consistent with $\sqrt{-g} = 1$. Since then, I have handled Newton's case differently according to the final theory.— Thus there are no gravitational waves analogous to light waves. This probably is also related to the one-sidedness of the sign of scalar T, incidentally. (Nonexistence of the "dipole".)
>
> Cordial greetings and many thanks for the interesting communication. Yours,
>
> A. Einstein.

In this expanded passage we see Einstein discussing the special choice of coordinates he found useful to make when treating gravitational waves. General relativity takes its name from the fact that the same set of equations govern the behavior, in a gravitational field, of any object, no matter how it moves and no matter how the observer moves. In other words, no matter what your point of view—and as Einstein already showed in his 1905 paper on special relativity, how you measure time and space depends heavily on your point of view, or "state of motion"—the same set of equations describes the physics you experience. Of course, if you want to use these equations to calculate things that a real person would be able to measure, then, since the measurements depend on one's point of view, one is obliged to adopt a point of view in the calculations. This point of view is embodied in the geometrical coordinates chosen for the calculation. Although the Einstein equations are independent of any coordinate system, it is still necessary, when doing calculations with them, to adopt a definite coordinate system for a definite observable result. The choice of coordinates is very important. Although in principle any coordinate system may be chosen to suit the whim of the theorist, in practice only some will be appropriate to the problem at hand. If the coordinate system is ill chosen, a calculation may prove impossible to carry through. Worse, as we shall see, a poorly chosen coordinate system or frame of reference will simply cause one to get the wrong answer (if, for instance, the coordinate system you actually impose in your calculations does not in truth correspond to the observational point of view you intended it to represent). The coordinate system will, for instance, reflect whether the putative observer is moving with or without acceleration, or whether she is in a gravitational field or drifting in space.

In his earliest papers on general relativity, these from November 1915, Einstein used what are sometimes called unimodular coordinates, and he seems to have thought that these coordinates would prove to be so generally simplifying that they would represent the "usual" choice in calculations involving his theory. The idea was that a certain quantity, related to the determinant of the metric, would be set equal to unity, so that $\sqrt{-g} = 1$, which meant, for instance, that quantities called tensor densities were numerically equal to the tensors they were derived from. Presumably Einstein's hope was that this would greatly simplify many calculational steps.

Although Einstein and others had, at this time, a good conceptual grasp of the meaning of coordinate systems and their role in a general relativistic theory, they had as yet had little time to develop a strong intuitive feel for their use. Previous theories, such as that of Newtonian gravity, had postulated a unique coordinate system against which all others were referenced, although in practice many important questions had always turned on the choice of a frame of reference, as in the conflict between the Sun-centered system of Copernicus and the Earth-centered system of Ptolemy. Ironically, general relativity now claimed that the Earth-centered system was perfectly acceptable and even preferable for some practical purposes. One just had to be careful to remember that it was a choice one made to see the universe from that perspective.

It is perhaps not surprising that Einstein presumed that there might be one coordinate system which would simplify a wide range of calculations, and he advocated the unimodular system enthusiastically in his early papers. The paper of November 4, which he referred to, was his first paper of that month, and in it he presented a set of field equations close to, but different from, the final form the field equations took later in the same month (and which they have retained to this day). These field equations of November 4[1] behaved similarly enough to the true Einstein equations to permit some calculations to get the right answer, for instance, Einstein's calculation of the perihelion advance of Mercury. But to Schwarzschild Einstein noted that his then customary choice of unimodular coordinates was "no longer consistent" with another coordinate constraint mentioned in that paper, that in which $\frac{\partial g^{\mu\nu}}{\partial x_\nu} = 0$. This comment is interesting because the latter condition is similar to the one Einstein eventually used in his work on gravitational waves. But for the time being, in February 1916, he said that since discovering the final form of the field equations, he had "handled Newton's case differently," presumably a reference to his having to recalculate his Newtonian order approximations with the final field equations and to having to modify his coordinate choice. The result was that he convinced himself that "there are no gravitational waves analogous to light waves."

We cannot know from this remark exactly what convinced him that gravitational waves do not exist. If his calculation simply involved a post-Newtonian approximation, it is not at all surprising that he should have reached this conclusion, since a consistent Newtonian-order approximation of general relativity is not very well suited to describing gravitational waves. One must devote a lot of effort to seeing any evidence of the role of gravitational waves in such an approximation.

The important point here is that Einstein's theory is what is known as a metric theory: it employs a metric that describes how a given observer will make measurements of time and space. The metric, typically written $g_{\mu\nu}$, is a tensor that can be written as a matrix with four rows and four columns, or sixteen components in all. This tensor is used to build up one side of Einstein's equations, the left-hand side, which describes the curvature of spacetime (the quantity on this side of the equations is known as the Einstein tensor). Each component represents a separate equation,

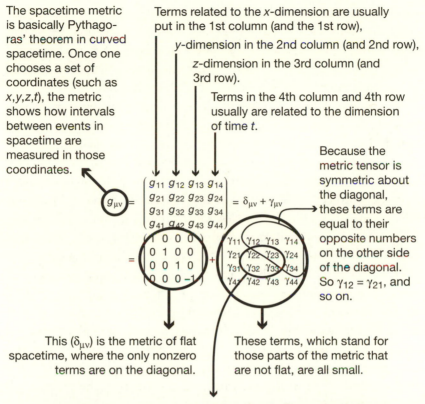

The spacetime metric is basically Pythagoras' theorem in curved spacetime. Once one chooses a set of coordinates (such as x,y,z,t), the metric shows how intervals between events in spacetime are measured in those coordinates.

Terms related to the x-dimension are usually put in the 1st column (and the 1st row), y-dimension in the 2nd column (and 2nd row), z-dimension in the 3rd column (and 3rd row).

Terms in the 4th column and 4th row usually are related to the dimension of time t.

Because the metric tensor is symmetric about the diagonal, these terms are equal to their opposite numbers on the other side of the diagonal. So $\gamma_{12} = \gamma_{21}$, and so on.

This ($\delta_{\mu\nu}$) is the metric of flat spacetime, where the only nonzero terms are on the diagonal.

These terms, which stand for those parts of the metric that are not flat, are all small.

If gravitational waves traveling in the x-direction are present then, in Einstein's coordinate choice, these terms (γ_{22}, $\gamma_{23} = \gamma_{32}$ and γ_{33}) must be varying with time in a periodic fashion. The waves are transverse because the motion is in the x-direction, while the oscillations are in the y-z plane (terms only in columns 2 and 3).

Figure 3.1. The linearized approximation of general relativity.

but out of the sixteen possible equations, six are duplicates because of the symmetry of the equations, leaving the ten distinct equations mentioned previously. A defining characteristic of tensors is that they can be transformed from any system of coordinates to any other system in a way that preserves certain important quantities known as invariants. All observers will agree on the values of these invariant quantities.

A conventional way of representing the components of the metric tensor is to identify one of its rows (and therefore, because of the symmetry mentioned earlier, one of its columns) with the dimension of time and the other three with the dimensions of space. This representation will be part of a coordinate choice, and one should remember that general relativity is so general that one can choose coordinate systems with two timelike dimensions and two spacelike dimensions, if one chooses.

When one employs a Newtonian or post-Newtonian approximation of general relativity, one loses this metric aspect of the theory. The reason is that the leading-order terms come from just one component of the Einstein equations. This is expected since Newtonian theory is what we would now call a scalar theory of gravity, in that the gravitational field can be described by a single number instead of by the ten or sixteen components of a tensor.[2] This means, of course, that the Newtonian approximation of general relativity does not retain the linkage of space and time characteristic of relativity theory, and while this presents no difficulty in describing most aspects of gravity, it does create a problem in describing gravitational waves as they travel far from the physical system that generated them. The gravitational field of the source system in the Newtonian approximation weakens according to the famous inverse-square rule, but in fact gravitational waves do not weaken as quickly as that when one is far from the source. As we shall see, the structure of spacetime far from any source is crucial to the representation of gravitational waves, and this structural aspect of the theory tends to be lost in the Newtonian approximation.

It is not impossible to find evidence for the existence of gravitational waves in the post-Newtonian approximation, though it requires much labor. When one reduces the Einstein equations to a solution expanded in powers of v/c, the terms that do not depend on v/c at all represent Newton's law of gravity.[3] Now if one looked for terms that would describe the effects of a finite speed of propagation of gravity, as Laplace did, one might expect to find terms of order v/c, as he did, but in fact there are no such corrections to Newton's results in general relativity. The first post-Newtonian corrections are terms containing squared powers of v/c, that is to say, terms of order $(v/c)^2$, as in special relativity. What is significant is that terms with odd powers of v/c are, in general, as with Laplace's terms, nonconservative. They are associated with losses of energy by the material system, and nowadays we would expect that the lost energy is being carried away by gravitational radiation. Terms with even powers of v/c are conservative terms. So, for instance, the term describing the perihelion advance of Mercury is of order $(v/c)^2$ and is therefore a conservative term. This means that although this term revolves the orbit of Mercury slowly about the Sun, it does not cause Mercury to drift closer to the Sun with time and ultimately fall into it, as the loss of energy to gravitational wave emission would do. The next order would be terms containing three powers of v/c, but again there are no terms of order $(v/c)^3$ in the general relativistic post-Newtonian approximation. After some more conservative terms of order $(v/c)^4$ (referred to as second post-Newtonian order because of the habit of counting in units of $(v/c)^2$), one finally finds nonconservative terms indicating the effects of gravitational radiation only at order $(v/c)^5$. It is certain that Einstein did not go so far in his calculations in 1916. Therefore if he did look for nonconservative terms of the Laplace type, he would certainly have noticed that they did not exist at low order. This would have been of particular interest to him, because as Poincaré had shown, such terms would probably affect the perihelion advance of Mercury, since as Mercury spiraled slowly in towards the Sun, there would most likely be an attendant shift in its perihelion. Since the wonderfully precise agreement between general relativity's prediction of the perihelion advance of Mercury and the astronomer's observations was a key reason why Einstein and others believed in the correctness of the theory, Einstein might have paid quite close attention to any terms that would have affected it. So it is possible, though purely speculative, that from the absence of any nonconservative perturbation to the motion of Mercury Einstein concluded Mercury could not be emitting gravitational waves, which therefore most probably did not exist. As it turns out, Mercury moves much too slowly to exhibit any measurable effects of gravitational radiation.

Einstein seems to have guessed from this that the absence of gravitational radiation from the theory might be connected to one obvious breakdown in the analogy with the electromagnetic force: the nonexistence of a push companion to the gravitational pull; that is to say, the fact that no negative masses exist which might produce an antigravity effect on ordinary mass.

Therefore, we see that the question of whether gravitational waves existed was one of the first questions addressed by Einstein after he completed his theory. In this we can see the suggestiveness of the analogy with the electromagnetic field, but we also see that Einstein's response was skeptical. Gravitational waves do not exist, he decided, and he speculated that the reason lay in the incompleteness of the analogy between the two field theories.

Within a few months of reaching this conclusion, Einstein was to completely reverse his opinion. The occasion was his development of another approximation scheme, which was based on special relativity rather than on Newtonian theory. As it turned out, that scheme was suited to describe gravitational waves in a way that the Newtonian one was not. Einstein's discovery that gravitational waves were inherent to such a scheme, which is known as the linearized approximation scheme, was made possible by a suggestion from the astronomer Willem de Sitter. De Sitter, one of several Dutch colleagues of Einstein who took a strong early interest in his new theory, made a number of important contributions to the theory of general relativity in its early days, especially in the area of cosmology. In fact the two men's exchange of letters on the linearized approximation occurred at the beginning of a long argument between them on the subject of Einstein's cosmological ideas. De Sitter's greatest contribution to the history of relativity is probably that his papers on the subject which were published in an English astronomy journal in 1916 introduced the theory to English astronomers at a time when the ongoing First World War prevented any direct contact between English and German scientists. De Sitter thus contributed to the dramatic vindication of relativity theory that occurred in 1919 when an English expedition verified Einstein's prediction that starlight would be deflected from its path by the gravitational field of the Sun (see Stanley 2003 for an account of the expedition and Warwick 2005 for de Sitter's role in transmitting details of general relativity to Britain).

The idea behind the linearized approximation of general relativity is to find an approximation scheme that preserves the metric quality of the theory. Special relativity can also be understood as a metric theory, but one in which the metric represents flat, or Euclidean, spacetime. As we know in general relativity, one treats spacetime as curved in the presence of gravity. But if gravity is weak, then the amount of curvature is small, and indeed, in our everyday experience here on Earth, we know that spacetime is very, very close to flatness. Since approximation schemes require a small quantity in powers of which the expansion can be made, it makes sense to write the general relativistic metric $g_{\mu\nu}$ as the sum of a flat, special relativistic metric ($\delta_{\mu\nu}$) and a metric ($\gamma_{\mu\nu}$) whose components are all very small (therefore $g_{\mu\nu}$ is close to flatness). We can then write the Einstein equations in powers of $\gamma_{\mu\nu}$. Terms in which only $\delta_{\mu\nu}$ appear will agree with the results of special relativity, and the first-order corrections to this will be of first or linear order in $\gamma_{\mu\nu}$. Since one of the great difficulties with full general relativity is that it is nonlinear, equations involving only linear (i.e., first-order) powers of $\gamma_{\mu\nu}$ are much easier to solve. Hence the name linearized approximation for this approach. In mid-1916 Einstein devoted a paper to his presentation of the linearized approximation in which he introduced gravitational waves to the theory of general relativity.

The linearized approximation is not as well suited as the Newtonian approximation for treating the orbital motion of planets, but it does have the great advantage that it permits the Einstein equations to be written in a form which is closely analogous to the way that Maxwell's equations for the electromagnetic force are written in the theory of special relativity. Because one of the difficulties with a new theory is that it takes time to establish the intuition which guides difficult calculations, it is always helpful to recast the equations in a more familiar, even if approximate, form. A clever choice of coordinate system aided Einstein in reducing the equations to the desired form. It seems that to begin with, Einstein continued using his unimodular coordinates while working with the linearized equations and did not make much headway; but on June 22, 1916, he wrote gratefully to de Sitter, having made a breakthrough:

> Highly esteemed Colleague,
>
> Your letter pleased me very much and inspired me tremendously. For I found that the gravitation equations in first approximation can be solved exactly by means of retarded potentials, if the condition for the coordinate choice $\sqrt{-g} = 1$ is abandoned. Your solution for the mass-point is then the result upon specialization to this case. Obviously your solution differs from my old one in the choice of coordinate system, but not intrinsically.
>
> (Einstein 1998, Doc. 227)

Here Einstein thanked de Sitter for showing him that a different choice of coordinates would allow him to solve the linearized equations of gravity without any further approximation (as he put it, "exactly"). What is more, he could solve these equations by means of retarded potentials, which play a central role in the theory of electromagnetic radiation and will play an equally important role in our story. *Potential* is, of course, simply another word for the field produced by a source, referred to by Einstein here as the mass point. If we suppose the source is moving, how should we describe the field or potential it produces at a distant point? Should it be the field produced from the position the source currently occupies? If so, we presume that the field propagates instantaneously, and we have an action-at-a-distance theory. If we presume that it takes some finite time for any changes in the field to propagate from the source to the distant point, then we must presume, as Laplace did in his calculation in the pre-field-theory era, that the field or force at the distant point must appear to come from the position occupied by the source at the time the field, as it were, set out. This position is called the retarded position. The field it produces is called the retarded field, or the retarded potential, and the time at which the field was emitted is called the retarded time. If we can solve for the retarded position of a massive body, then we can deal with the case when the massive body is moving, even accelerating. Keep in mind that Schwarzschild's solution for the gravitational field of a massive body applies only in the case where the body is at rest. The motion of Mercury can be dealt with only in the approximate case where it is presumed to be moving slowly around the stationary Sun. But we would like to deal with the case of a body moving or accelerating fast enough to produce significant gravitational waves. To do this, we would really like to be able to solve the equations in terms of the retarded potential.

In his paper on the linearized approximation, published in mid-1916, Einstein succinctly described what had been achieved:

> We shall show that these $\gamma_{\mu\nu}$ [the small perturbative quantities] can be calculated in a manner analogous to that of retarded potentials in electrodynamics. From this it follows that gravitational fields prop-agate with the speed of light. Subsequent to this general solution we shall investigate gravitational waves and how they originate. It turned out that my suggested choice of a system of reference . . . is not advantageous for the calculation of fields in first approximation. A letter note from the astronomer De Sitter alerted me to his finding that a choice of reference system, different from the one I had pre-viously given [in 1915, p. 833], leads to a simpler expression of the gravitational field of a point mass at rest. (Einstein 1916, p. 688)

So the analogy now passed from being merely suggestive to a precise cor-respondence between two sets of equations, one being the vector equations describing the electromagnetic field, the other being the tensorial equations describing the gravitational field in linearized approximation. Employing the same procedure that one would in the electromagnetic case, it naturally transpires that both sets of equations lead to a wave equation. If the electromagnetic wave equation describes the well-known and experimentally well-understood electromagnetic waves, it seems reasonable to conclude that the gravitational wave equation describes gravitational waves, which propagate at the same speed as light but which consist of oscillations in the gravitational field.

Einstein derived solutions to his wave equation that were intended to represent plane gravitational waves, by which we mean waves whose wave front is flat, rather than curved. They can be thought of as waves that are so far from their source their spherical shell, centered on the source, appears flat at any one point, just as the Earth appears flat when one is standing on it. Einstein proceeded to classify the waves into three types representing different symmetries. He called these types longitudinal, transversal (referring to transverse and longitudinal waves familiar from waves in other media), and a "new type of symmetry," arising from the third set of conditions.

Now it would be interesting to know whether the waves carry energy. Employing a pseudo-tensor quantity derived in the first part of the paper, Einstein proceeded to calculate the energy transported by these waves. He found that "only waves of the last named type [the "new type" of wave] do transport energy." What could this mean? Did it make any kind of sense to speak of a wave that did not actually carry any energy along with it? How would one even know it existed? This obviously puzzled Einstein, and he discussed the problem with de Sitter (continuing the letter quoted above):

> As it is, one could think that the coordinate choice $\sqrt{-g} = 1$ were not at all natural. However, I have found a very interesting physical justification for the latter. I denote the $\sqrt{-g}$ system as K, the generalized de Sitter system as K'. We now inquire about plane gravitational waves.

So anything we see in system K is seen from the vantage point expressed by Einstein's previously preferred unimodular coordinates. System K' refers to the choice of coordinate used by de Sitter, which Einstein found so useful from a calculational point of view. These coordinates, of which more will be said below, are sometimes referred to as isotropic coordinates, since the velocity of light is the same in all directions in them (in general relativity, the velocity of light is not constant, as we see from the well-known result that light follows a curved path in a gravitational field). Einstein continued:

> In system K' I find 3 types of waves, of which only one is connected to energy transportation, however. In system K, by contrast, only this energy-carrying type is present. What does this mean? This means that the first two types of waves obtained with K' do not exist in reality but are simulated by the coordinate system's wavelike motions against Galilean space. The ($\sqrt{-g} = 1$) system thus excludes undulatorily moving reference systems, which simulate energy-less gravitational waves. System K' is nonetheless useful for the integration of the field equations in first approximation. I am curious about your Moon paper and am anyway delighted that you take such pleasure in the general theory of relativity.
>
> Cordial regards to you, yours very truly,
>
> A. Einstein.

So we begin to see how tricky a business is this choice of coordinates. De Sitter's isotropic coordinates are preferable for actually doing the calculation, but then it seems they can lead one astray, because they simulate "wavelike motions" that make it appear a wave is present in spacetime when there is none. In other words, someone observing from such a frame of reference would observe periodic changes in the distances between objects that would appear characteristic of a gravitational wave but that would actually be the result of the observer's own noninertial periodic motions. Another observer might see no such oscillations, but they both agree on the apparent existence of the third type of wave, and since this type is also the one which appears to carry energy, Einstein concluded that it is the real wave and the other two types are merely apparent. Here the effort to follow through the analogy with electromagnetism is shown to have its potential pitfalls, because one is obliged to introduce "undulatorily moving reference systems, which simulate energy-less gravitational waves."

So if gravitational waves are, as people later came to say, "ripples in the curvature of spacetime," how is it that flat space, with no ripples, can appear to be wavelike when a certain coordinate system is used? If it is unclear how flat space can be made to appear wavy merely by the introduction of a system of coordinates, consider the following scenario, in which you play the role of a marine scientist interested in the locomotive properties of seaweed and other marine algae.

Suppose you have a tank of seawater at your lab with some seaweed in it. You will notice that the seaweed never moves, since there are no currents in the tank. It seems that seaweed, having no muscles, is inert and incapable of independent motion. Appearances can be deceiving, so to confirm your hypothesis, you decide to make some measurements. After all, science is all about measurement, which, being objective, never fools the observer. You take a rod and measure the distance from the seaweed to the side of the tank every second. It never varies, convincing you that your immediate sensory impression was not mistaken.

Now suppose you go to the seaside, where you notice that seaweed hanging in the water near the shore perpetually moves back and forth, making the same motion over and over again. Does seaweed in its natural habitat possess the power of motion? It is not hard to figure out that the apparent motion of the seaweed is in fact due to the periodic motion of the water in which it hangs. In fact, the seaweed is a good visual clue to the presence of waves in the water. Although water waves are easy to discern on the surface of the water, if one looks at a volume of water beneath the surface, it would only be the motion of seaweed or other objects suspended in the water that would inform you the water was moving (assuming you were not also moving with the water).

You decide to perform a series of measurements. First, you place a cork on the surface of the water and measure its distance form the shore every second. At zero seconds it is 10 feet from the shore. After a second it is 9 feet. Another second finds it at 8 feet, but after three seconds, it is back to 9 feet, and at four seconds it is at 10 feet, and so on, in an unvarying rhythm. Since the motion of the cork is presumably entirely due to the motion of the water, it seems logical to suppose that the seaweed is also still unmoving with respect to the water it hangs in. To confirm this you observe the distance between the seaweed and the cork. Sure enough, the seaweed never changes its position with respect to the cork. It is apparent that, with respect to the water it floats in, the seaweed in the ocean does not move any more than does the seaweed in the tank; it merely partakes in the wavelike motion of the water. This local measurement of the motion of the seaweed with respect to the cork is an excellent method of detecting the presence of waves, unless, of course, the distance between seaweed and

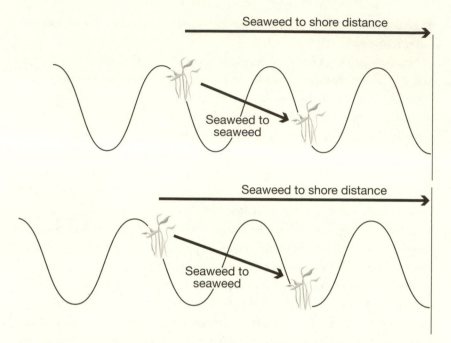

Figure 3.2. Seaweed waves. As the seaweed moves in the waves, the distance between a piece of seaweed and the shore changes, and the distance between two pieces of seaweed changes. Either change is an indication of the presence of waves. Far from land, the distance from shore measurement can still in principle be used, provided one has a coordinate system that covers the whole ocean. But one is increasingly prone to errors, because one cannot actually see the coordinates. In addition, many different kinds of coordinate systems could be chosen, and one must be careful to work consistently with a given system to avoid confusion.

cork is much less than the wavelength of the wave. In that case they will move in concert with each other at all times, as they react to the same part of the wave motion up and down.

You decide to make a new set of measurements in the seaweed's wave coordinates. To do so, you measure the distance to the shore every second, but at one second you add 1 foot to the total; at two seconds you add 2 feet; at three seconds, 1 foot; at four seconds you add nothing; and then at five seconds, 1 foot again, and so on. The upshot will be that if the seaweed is only moving with the water you will again always get a corrected measure of 10 feet from the shore, completely unvarying over time. This method, measuring the seaweed's motion with respect to the shore, is not a great way to detect the presence of waves if one is out of sight of land, but a shore-oriented coordinate system is, of course, exactly what one would like to have in mid-ocean when trying to find land, and the whole purpose of the longitude work discussed in chapter 2 was to provide seamen with a means of plotting their position on such a coordinate system.

But now suppose you return to your waveless tank and decide to have your assistant repeat your first experiment, just to be sure. Let us suppose the assistant is confused and thinks you wanted to use the new "corrected" way of measuring that was used at the beach. Accordingly, what the assistant does is measure the position of the seaweed in the tank every second, and then add 1 foot to its position after the first second, two feet after the second second, 1 foot after the third second, and so on. In other words, the wave coordinates are used, when there is in fact no wave at all present. The assistant leaves the list of measures on your desk, and upon your return in the morning, you are astonished at what they contain. Instead of an unvarying position typical of seaweed in the tank, you find a periodic variation, first 10 feet, then 11, then 12, then 11, and so on. Are there actually waves in your tank you had never noticed? Your assistant may be a bit dim but is usually very reliable with a measuring rod.

You go over to the tank and look in. There is absolutely no sign of any waves. The surface of the water is flat, there is no sound of waves or ripples slapping off the walls of the tank, and most of all, the seaweed does not appear to be moving at all. Are appearances deceiving you? You take out your measuring rod and satisfy yourself that the seaweed remains always 10 feet from the side of the tank. After poring over the assistant's figures for a while you realize what has happened. He has used the wave coordinates, thus making it appear, on a straightforward reading of the figures, as if there had been a wave in a tank when there was none! It turns out that it is not just appearances which can be deceiving, so can measurements. In fact, measurements are very far from being objective. In reality, as Einstein realized first in 1905, nothing is more subjective than measurements, because the existence of a measurement presumes the existence of a point of view, or frame of reference, and most measurements are meaningless without a coordinate system that defines them. In this case the assistant came close to convincing you that there was a wave in your tank when its surface was as flat as a pancake. But what about waves that exist in spacetime itself? How can we easily see whether spacetime is flat, especially when we have never measured how it behaves when it is waving? This was to be the challenge of gravitational wave research.

In an appendix (*Nachtrag*) added to the end of his 1916 paper (suggesting that Einstein only thought of this solution to the problem of the energy-less gravitational waves after presenting his paper in person to the Prussian academy), Einstein returned to the

> strange result that gravitational waves which transport no energy, could exist. The reason is that they are not "real" waves but rather "apparent" waves, initiated by the use of a system of reference whose origin of coordinates is subject to wave-like jitters. (p. 696)

Referring to the question of which of the two coordinate choices is preferable, he offered the following closing remark in the appendix:

> Even though it turned out to be advantageous for this investigation not to restrain the choice of coordinate system for the calculation of a first approximation, our last result shows nevertheless that the choice of coordinates under the ["unimodular"] restriction has a deep-seated physical justification. (p. 696)

Einstein later clarified his position on the "deep-seated physical justification" of these coordinates in correspondence with his colleague Gustav Mie. Mie was one of the few physicists (and the majority of them, like Mie, were German) who had been pursuing a unified field theoretic approach in the years leading up to 1916. He could appreciate general relativity's virtues better than most and could also be critical on a higher level than most physicists of the day could manage. He was largely unimpressed with the theory's claim to be a real "general" theory of relativity. For Mie there was still a definite distinction between inertial (unaccelerated) motion and noninertial (accelerated) motion that was inherent in the theory, despite its formal "covariance" (the property, mentioned earlier, that the field equations retained the same form whatever the observer's frame of reference or state of motion). Einstein's conviction was that there was no absolute state of motion; all motion was measurable only relative to the motion of other material bodies. When we are in a state of unaccelerated motion we are always free to presume that we are at rest: this is the axiom that Einstein called his principle of relativity. The principle of equivalence played just as fundamental a role in the general theory because it states that anyone in an accelerated state of motion is free to presume that they are really at rest in a gravitational field of the appropriate strength (just as when a pilot or astronaut experiences extreme centrifugal forces, he or she will refer to "pulling g-forces," meaning some multiple of the acceleration that gravity exerts on bodies at the surface of the Earth). But, says Mie, for a physicist there is one obvious difference between the state of acceleration and that of being in a gravitational field. In the former case the body will produce radiation: electromagnetic waves if it is charged, gravitational waves regardless; but in the latter case, where the body is at rest, it will not. As we shall see, this subtle point was to resurface later in the debate over the existence of gravitational waves. If accelerating bodies radiate, then when in a theory such as general relativity are we to understand that a body is accelerating?

To Mie, therefore, general relativity was not all it was cracked up to be, as far as some of the more "philosophical" claims made for it went. On the other hand he was impressed with its success as a physical theory. To Mie's way of thinking, a coordinate system in which a perfectly straight rod appeared to be "writhing" (or "slithering") simply could not to be considered as useful as one in which the rod looked straight. Surely the coordinates in which a straight rod was simply a straight rod were more real than those in which spacetime itself never stopped writhing. It was a further claim of Mie's that some coordinate systems give rise to gravitational or other fields that appear to have no causal explanation, even though such fields do not lead to any unexplained behavior of material systems. We ourselves may observe that since gravitational waves are composed of pure field, an apparent gravitational wave, such as those which appear in Einstein's 1916 paper, might be considered an example of a field that has no source or other causal explanation and that does not affect matter since it does not transport energy anywhere.

In a long letter to Einstein on February 5, 1918, Mie noted that he had held up the writhing rod as a "frightful exemplum" of the absurdities of general covariance. Einstein replied on February 8:

> With your considerations on the curved (fluttering) rod I *completely* disagree. All physical descriptions which lead to the same observable consequences (coincidences) are fundamentally equivalent, provided both descriptions are based on the same physical laws. The choice of coordinates can be of great practical importance from the point

of view of the clarity of the description; but in essence it is totally meaningless. It doesn't mean anything that depending on one's choice of coordinates there are "arbitrary gravitational fields;" the fields themselves cannot lay claim to reality. They are only analytical auxiliaries to describe realities. In fact, one really only finds out about the latter through elimination of the coordinates. The ghost of absolute space haunts your rod example.(Einstein 1998, Doc. 460; this translation by Michel Janssen)

As Einstein emphasized here, it is an important point in modern relativistic field theories that the fields themselves are not actually observable. It is quite possible to do a calculation involving fields in several different ways, and in each of them, the quantities describing the field and quantities relating to it, such as the amount of energy in the field for instance, will be vastly discordant. But in each case the description of the physical sequence of events relating to material bodies will be in agreement. The different ways of doing the calculation will involve different choices of systems of coordinates, or what are called gauge choices in most other field theories, such as electromagnetism. As Einstein said, the object of a gauge or coordinate choice is to make the calculation proceed more easily. Hence Einstein adopted de Sitter's suggestion concerning the isotropic coordinates, even though the results produced were difficult to interpret. The fact that one could easily mistake ordinary spacetime for gravitational waves in such a coordinate system shows that however useful such a system may be, it is still prone to misinterpretation, because it is only one way of looking at the world. This is particularly problematic when dealing with entities that are purely constructed of fields, as gravitational waves are. In one sense, as Einstein observed, we can say that fields are not "real," since by a mere mathematical reformulation we can make them behave anyway we want. But we are not totally free to disregard them, because the way in which they interact with matter is completely determined by the physics of the situation and must be essentially agreed upon by all observers. Gravitational waves can be made to do all sorts of strange things by a choice of coordinates, but the way they interact with matter must always make sense physically. The fact that the coordinate choice suggested by de Sitter is still routinely used in gravitational wave theory suggests that Mie's instinct that "slithering spacetimes" could not play a real role in physics was mistaken.

(De Sitter's coordinate choice is now called the de Donder gauge, and there is an even more important generalization of it called the harmonic gauge, which applies to the full nonlinear theory.)

The next episode concerning Einstein's 1916 paper illustrates a number of the themes already discussed in this chapter and prepares us for many of the debates that are to come. It turned out that Einstein made a severe error in this paper on gravitational waves, and it took him two years to correct it. The story of how he came to fix it was deciphered by my colleague and predecessor at the Einstein Papers Project, Michel Janssen. It involves the Finnish physicist Gunnar Nordström, one of the earlier pioneers of relativity theory, who was, in fact, the first to develop a fully self-consistent relativistic theory of gravity, a year before Einstein, although he was helped greatly in his work by Einstein himself, and it was Einstein who put Nordström's theory in the mathematical form in which it is known today.

In his paper Einstein wished to calculate how much energy would be carried by gravitational waves from a material source. To do so, he needed a way to describe the amount of energy in a gravitational field. We have just discussed how field quantities vary according to one's choice of coordinate system, and gravitational field energy is no exception. Einstein recognized this very early on, so that although he described the stress energy of material things by a tensor, thus making the total stress energy of a system an invariant quantity which would be unchanged by coordinate transformations, he chose to describe the energy in the gravitational field by a quantity known as a pseudo-tensor: it looks superficially like a tensor but lacks the marvelous property of invariance to all coordinate transformations. Several relativists in the early days of relativity criticized Einstein for this aspect of the theory (it was highly controversial), but Einstein was unmoved by the objections. He realized that the equivalence principle demanded that if we imagine that the house we live in is a really a rocket ship accelerating through space, then it is always possible to treat a gravitational field as if it were simply an accelerating system. This result suggests that if a gravitational field can always be made to disappear (at least at a given location) by the correct choice of coordinates, then the energy in that field must disappear also (though it must, of course, reappear somewhere else under another guise, in order to satisfy energy conservation for the whole system in question). Since the energy in gravitational waves would have to be described by these pseudo-tensors, it was always going to be problematic. But for now, Einstein's problems were more straightforward: he just blundered while calculating the form of his pseudo-tensor in his 1916 paper.

It will be recalled that Einstein's scheme in this paper involved writing the metric of spacetime as the sum of the flat-space metric and a metric, written $\gamma_{\mu\nu}$, whose components are small and which represents small departures from curvature in spacetime due to the gravitational field. He immediately introduced a new set of quantities, labeled $\gamma'_{\mu\nu}$, which depend on the γs. It is on these γ' quantities that the isotropic coordinate system is imposed, and it is these quantities that are treated as the gravitational field in analogy with the usual electromagnetic vector potential. Therefore the gravitational waves themselves will be described in terms of these convenient γ's. However when Einstein defined his pseudo-tensor, he wrongly defined it in terms of his γ' quantities instead of his original γs, which actually describe the metric, which means that the pseudo-tensor he derived was not correct. He was fortunate in that some of the conclusions he drew were actually true in spite of his error. For instance, had he derived the pseudo-tensor properly, he would still have found that the two spurious waves did not transport energy (though as we shall see, this is by no means a definitive argument against their reality). But when he derived an expression for the amount of energy radiated by a material source of gravitational waves, he recovered a hopelessly wrong formula. The most striking quality of the radiation formula derived in the 1916 paper is that it permits the radiation of energy from monopole sources.

Now Abraham had previously recognized that fundamental considerations of physics demanded that gravitational waves could not be emitted by sources displaying motions with only monopole or dipole symmetries (so for instance, a balloon that expanded and contracted while making a spherical shape the whole time would be an example of a monopole source, while a perfectly spherical planet rotating on one axis would be an example of a dipole source). Therefore, the fact that Einstein's formula permits such radiation would make Abraham or any modern relativist suspicious. But apparently Einstein saw no difficulty. If we look back to his letter to Schwarzschild, we see that he himself wondered out loud whether the absence of a gravitational dipole could be connected to the "non-existence" of gravitational waves, but it seems that he must never have followed this line of argument much beyond this tentative speculation; otherwise, he might have realized right away that he had erred in his 1916 linearized calculation. Instead it was not until Nordström wrote to him the following year that he first had an inkling that something was amiss.

Nordström was interested in the question of how to calculate the mass of a body in Einstein's theory. It might seem odd that there should be any difficulty in calculating a quantity that is as fundamental to a theory of gravitation as the mass of a body, but it a good indication of the subtleties of this kind of field theory that it has been a subject which has given relativists much trouble. General relativity is a nonlinear theory because if the gravitational field contains energy, then it also has mass (recall the most famous equation of relativity theory, $E = mc^2$) and so it must itself generate a gravitational field. Therefore, doubling the mass of a body will increase the strength of its gravitational field by more than a factor of two, because the larger field itself generates an attractive force on other bodies. What this means is that, depending on how you set the problem up, if you want to calculate the mass of a body you must not only sum up (i.e., integrate) all the material in the body itself but must also integrate over all space to include the energy in its gravitational field. Nordström discovered that there was a way to arrange matters so that the mass of the body could be calculated by an integration extending only over the space occupied by the body itself. The natural result of this idea, as he outlined in detail in a letter to Einstein of September 22, 1917, is that there is no energy whatsoever stored in the gravitational field permeating all of spacetime surrounding the body. This seems a rather curious result, and so Nordström decided to check it. A useful tool to hand was the pseudo-tensor derived by Einstein in his gravitational wave paper. It can certainly be used to calculate the field energy of a stationary source. But the result it provides is nonzero for the field energy.

In the midst of writing his letter to Einstein raising this discrepancy, it occured to Nordström what the source of the problem must be. He was working in a different coordinate system from the one used in Einstein's paper. Nordström used the unimodular coordinates adopted by Einstein in most of his early papers (Nordström follows one paper in particular, also of 1916), whereas the pseudo-tensor derived in the gravitational wave paper used the isotropic coordinates of that paper. Redoing his own calculations in the light of this realization, Nordström did indeed calculate a nonzero result for the gravitational field of the source, but one which still disagreed, by a factor of two, with the result obtained by using Einstein's pseudo-tensor

The Einstein tensor is built up from the metric tensor and the Riemann tensor (which describes the curvature of spacetime).

The other side of the Einstein equations contains the stress energy tensor.

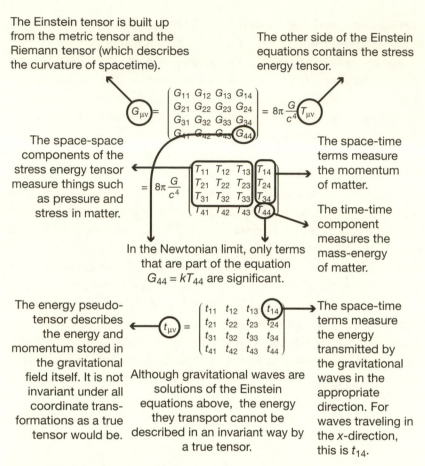

The space-space components of the stress energy tensor measure things such as pressure and stress in matter.

The space-time terms measure the momentum of matter.

The time-time component measures the mass-energy of matter.

In the Newtonian limit, only terms that are part of the equation $G_{44} = kT_{44}$ are significant.

The energy pseudo-tensor describes the energy and momentum stored in the gravitational field itself. It is not invariant under all coordinate transformations as a true tensor would be.

Although gravitational waves are solutions of the Einstein equations above, the energy they transport cannot be described in an invariant way by a true tensor.

The space-time terms measure the energy transmitted by the gravitational waves in the appropriate direction. For waves traveling in the x-direction, this is t_{14}.

Figure 3.3. The Einstein equations and the pseudo-tensor.

ostensibly applying to the same coordinate system. Thus Nordström was left with two questions, one of which was the quite general one of whether the pseudo-tensor approach to describing the energy in the gravitational field was really of any use ("I must say, though, that the interpretation of [the time-time component of the pseudo-tensor] as the energy density [of the gravitational field] does not seem as useful to me now as before"). The other questions was the specific one concerning the reason for the remaining discrepancy between Nordström's and Einstein's results ("Maybe this inconsistency is owing to an incorrect approximation").

Einstein's reply has not survived, but that he did make one is shown by Nordström's following letter of October 23. From this it appears that Einstein must have tried to replicate Nordström's calculation and protested that he did not find that the energy in the field of a stationary mass in unimodular coordinates was zero. Nordström replied, reiterating his own calculation, apparently convinced that Einstein must have made a mistake, as indeed he had. It appears that Einstein learned three lessons from this episode: first, confirming what he already believed, that presuming that one coordinate system gives a more trustworthy or "intuitive" picture of reality than others is dangerous; second, that using pseudo-tensors to show that there is no energy in a gravitational field does not mean that that gravitational field "does not exist"; and third, that his pseudo-tensor and therefore his radiation formula from 1916 were completely incorrect. As we shall see, arguing that gravitational waves do or do not transmit energy solely on the basis of pseudo-tensor calculations is inherently dubious.

Although we have no other letters in the exchange, Einstein obviously picked up on Nordström's point pretty quickly after the second (October 23, 1917) letter, because by the end of January 1918 he had submitted a new paper that repeated the main points of the 1916 paper but corrected the mistake in the pseudo-tensor derivation. The main result of this paper, the celebrated quadrupole formula, plays a central role in the history of gravitational waves. This formula was the revised radiation formula, which expressed the amount of energy radiated by a source of gravitational waves. It gets its name, obviously, because the formula depends on changes in the quadrupole moment of the source. Thus it was confirmed that the lowest order of multipole emission in gravitational wave theory is quadrupole. Higher orders, such as octopole, are possible, though not considered by Einstein, but lower-order multipolar radiation does not exist. Disagreements over the correctness or applicability of this formula were to lie at the heart of most of the disputes on gravitational radiation over the next half century and more, but as we have seen, it was pure serendipity that permitted Einstein to correct the error of his 1916 paper and discover the quadrupole formula.[4]

4

The Speed of Thought

In his 1918 paper "On Gravitational Waves," Einstein reiterated the entire argument of his 1916 paper, with corrections, quite convinced of the usefulness of his isotropic coordinates for this problem. Once more he identified the three types of waves, two of them not transporting energy, even with the corrected pseudo-tensor. He then proceeded to convincingly demonstrate that, as he had indicated at the end of his 1916 paper, the non-energy-carrying waves were merely artifacts of his choice of coordinates. He showed that the metric for these waves can be arrived at by a simple coordinate shift from Minkowski flat space. Thus he compellingly illustrated that these "waves" are merely flat space seen in a curvy or writhing set of coordinates. Now there was no need to drag in the unimodular coordinates at all, and there was no need to rely merely on the fact that the pseudo-tensor indicates that these waves carry no energy (which, as Nordström's experience demonstrates, is no safe ground for presuming their nonexistence):

> Those gravitational waves which transport no energy can, there-fore, be generated from a field-free system by a mere coordinate transformation; their existence is (in this sense) only an apparent one. (p. 161)

We note here that Einstein's definition of a "field-free" system is one in which the metric corresponds to the "flat" metric of special relativity, known as the Minkowski metric. Thus, once again, we see that a gravitational field is identified with spacetime curvature. If there is no curvature, there can be no field and therefore no gravitational waves. We will see again later that this emphasis on curvature was to prove more reliable in describing gravitational waves than the tendency, inspired by the analogy with electromagnetism, to keep looking for the energy in the wave. Gravitational waves do transport energy, but it is always possible to represent that energy in different ways and in different places within a wavelength of the wave. In technical language, one says that gravitational field energy is nonlocalizable, a point which was recognized very early on in the development of the theory.

It might be thought that these spurious waves, which are really flat, had been decisively laid to rest. In practice they had quite a career ahead of them, and their persistence as ghosts haunting the de Donder gauge illustrates the pitfalls of choosing coordinates. Not only did they recur in the work of early relativists after 1916, but as late as the early 1960s, Richard Feynman (on p. 52 of the *Lectures on Gravitation*) introduced these same spurious waves, referring to them as "gauge waves," a term that reflected his own background in electromagnetic and quantum field theory. Apparently nearly everyone who works through linearized gravitational wave theory based on the analogy with electromagnetism "rediscovers" these nonexistent waves.

In 1919 Hermann Weyl introduced a section on gravitational waves into the third edition of his influential textbook on relativity, *Raum, Zeit, Materie*. He adopted a treatment that is clearly based on one or other of Einstein's two papers on gravitational waves. Following the argument of Einstein's linearized approximation closely, he arrived at the same decomposition of the field into three types of gravitational waves, which Weyl called longitudinal-longitudinal, longitudinal-transverse, and transverse-transverse. In short, we will refer to them as LL, LT, and TT waves, following Eddington (1922). The first two types correspond to the same spurious waves "discovered" by Einstein in 1916. The TT waves, also discovered by Einstein in 1916, are what nowadays are called gravitational waves, coincidentally still often referred to as TT waves, for the transverse traceless gauge they are commonly represented in. The strange thing is that Weyl simply ended his discussion at that point, leaving the LL and LT waves on the same footing as the TT waves. It seems that Weyl saw them as being just as real as the TT waves.

This is striking because Einstein stated quite clearly, of course, in the 1918 paper that what Weyl called the LL and LT waves are simply flat space and do not represent a wave solution of the linearized equations at all. Weyl must not have read this paper closely, for if he disagreed with

Einstein's conclusions, he was silent on the point. Obviously, having decided to adopt Einstein's approximation scheme in treating gravitational waves, Weyl did not closely follow the papers but simply developed the apparently straightforward argument for himself. Doing so inevitably led him to discover the same spurious waves, given the curious characteristics of the isotropic coordinate system, which lends spacetime a sinuous or "jittery" character, to use Einstein's description. Working quickly in the cursory fashion of a textbook, rather than in the analytical manner of a research paper, he must have failed to notice that there was something suspicious about these waves, as Einstein did when he attempted to calculate the energy they propagated.

How long this confusion might have persisted over the LL and LT waves is hard to say (Einstein himself still alluded to them in passing, before dismissing them for the third time, in his 1936 paper with Nathan Rosen, which will feature in chapter 5). Fortunately another prominent figure in the early development of general relativity next entered the fray: Arthur Stanley Eddington. Eddington was the leading astrophysicist of the day, the discoverer of the modern theory of the internal structure of stars. He was the popularizer of the theory of general relativity in the English language, and above all he was the leader of the English eclipse expedition to Africa in 1919 which had confirmed, in conjunction with another English expedition to South America, the most famous prediction of general relativity: that light itself would be deflected as it passed through the Sun's gravitational field. His 1922 textbook, *The Mathematical Theory of Relativity*, was the foremost English-language textbook for decades and had great influence on the field, because the middle of the twentieth century saw English becoming the language of science, as the United States of America took over the leadership role in the sciences.

Eddington's outlook on gravitational waves as expressed in his textbook has earned him a reputation as the first skeptic of gravity waves. As we shall see, that view accurately reflects his skeptical attitude towards the analogy with electromagnetic waves, although he was emphatically not one of those who concluded that gravitational waves did not exist.

Eddington was motivated by a concern over whether the analogy between gravitational and electromagnetic waves was really true in detail. He asked, do gravitational waves travel at the same speed as that of light? We have already discussed how physicists thought that, for the consistency of

Figure 4.1. Arthur Stanley Eddington, one of the founders of modern astrophysics. His main influence on the development of general relativity was through his celebrated textbook, in which his skeptical views on gravitational waves shone through. (Courtesy AIP Emilio Segré Visual Archives, Segré collection)

physics, the speed of propagation of gravity should be the same as, or at any rate not greater than, that found in Maxwell's theory. That idea had been one contributor to the emerging conviction that gravitational waves must exist. Therefore, Eddington reasoned, Einstein had tended to presume that gravitational waves would propagate at speed c. In his textbook Eddington followed Einstein's argument quite closely (without repeating Weyl's mistake or even mentioning the "spurious" waves) but expressed his dissatisfaction with Einstein's method, which seemed to him to presume the speed of gravity in advance:

> The statement that in the relativity theory gravitational waves are propagated with the speed of light has, I believe, been based entirely on the foregoing investigation [that is to say basically on Einstein's presentation of the problem]; but it will be seen that it is only true in a conventional sense. If coordinates are chosen so as to satisfy a certain condition which has no very clear geometrical importance, the speed is that of light; if the coordinates are slightly different the speed is altogether different from that of light. The result stands or falls by the choice of coordinates and, so far as can be judged, the coordinates here used were purposely introduced in order to obtain the simplification which results from representing the propagation as occurring with the speed of light. The argument thus follows a vicious circle. (Eddington 1923b, pp. 130–131)

Eddington then went on to propose that an examination of how spacetime curvature itself propagates through space might be a more illuminating way to proceed. He concluded "There does not seem to be any grave difficulty in treating this problem; and it deserves investigation."

Eddington took up his own challenge and wrote two papers on the subject of gravitational waves in the following years. In the first of these papers he looked at the emission and propagation of the waves themselves, and subsequently, in the second paper, at the question of the back reaction on the source, or the "problem of motion" of the source. In doing so, he identified keys points which would become major battlegrounds in the controversies marking the subject decades later.

Here is how he began the first paper, in 1922:

> The problem of the propagation of disturbance of the gravitational field has been investigated by Einstein in 1916, and again in 1918. It has usually been inferred from his discussion that a change in the distribution of matter produces gravitational effects which are propagated at the speed of light; but I think Einstein really left the

question of the speed of propagation rather indefinite. His analysis shows how the co-ordinates must be chosen if it is desired to represent the gravitational potentials as propagated with the speed of light; but there is nothing to indicate that the speed of light appears in the problem, except as the result of this arbitrary choice. So far as I know, the propagation of the absolute physical condition—the altered curvature of space-time—has not been hitherto discussed. (Eddington 1922, p. 268)

We note here that from the outset Eddington focused on the spacetime curvature of the waves (what he called the "absolute physical condition") as the surest guide to the physical behavior of the wave. Einstein's purpose in choosing to work in harmonic coordinates was to cast the linearized gravitational field equations in a form which is very close to standard formulations for the field equations governing electrodynamics. After doing so, it was not unnatural, according to Eddington, that he discovered wave phenomena moving with the speed of light. Eddington wondered whether Einstein had presumed the analogy with electromagnetic waves in order to prove it. To check one aspect of this, Eddington followed Einstein's general prescription for the harmonic gauge, while explicitly retaining as a parameter the speed of propagation, V, in order to relax Einstein's assumptions somewhat. Doing so gave him several conditions on the solution to the linearized equations that corresponded to the conditions describing the three types of waves derived by Einstein and again by Weyl, but Eddington's versions featured the extra parameter V. He then asked whether these waves really propagate with speed of unity (i.e., with a speed equal to that of light, whose speed is conventionally set equal to one by relativists, so that time and space are measured in the same units). The condition he derived for the TT waves is identical to those found by Einstein and Weyl, with the addition of the new condition that $V^2 = 1$. He also recovered the Einstein and Weyl conditions for the LT waves and for the LL waves, but in both cases with the addition of the V parameter in such a way that it is dependent on the metric components describing the "wave." But since the value of these components can be changed merely by a change of coordinates, V can be altered arbitrarily simply by a coordinate change. Only in the case of the TT wave is the condition on V ($V^2 = 1$) separate from the coordinate condition, so that the velocity has some independent meaning.

As Eddington famously put it, the LL and LT waves "are merely sinuosities in the coordinate system, and the only speed of propagation relevant to them is 'the speed of thought'" (p. 269).

Contemporary relativists often class Eddington with the later skeptics because of his famous remark about the speed of thought. One ought to keep in mind, though, that Eddington simply disproved the existence of certain spurious types of waves whose existence is firmly rejected by modern orthodoxy. He did not claim that all gravitational waves travel with the speed of thought. In the context of our current story, however, I think that Eddington can be seen as a model of the skeptic of gravitational waves. He proceeded from a position which was frankly skeptical of the a priori assumption that such phenomena exist and behave like their electromagnetic analogy. He began from an attitude of disbelief and took nothing for granted as he proceeded. His remarks towards the end of the paper on the quadrupole formula, which he rederived in this paper but corrected for a small, factor-of-two error in Einstein's formula of 1918, still expressed caution on whether the linearized analysis could be applicable to astronomical systems. He calculated the decay rate of atoms and binary stars due to the quadrupole formula, but noted:

> But both applications are probably illegitimate—the atom, because it is complicated by quantum conditions outside our analysis; the star, because the tension (being now a gravitational force) is limited to a small quantity of the first order, and the problem is thus carried out of range of our approximation. But it seems likely that the radiation (if any) will not exceed that given by [the quadrupole formula].
>
> There is clearly no practical objection to the existence of this small radiation from rotating systems, and I can see no theoretical reason for not admitting it. (p. 280)

It is important to realize that the skeptics do not reject outright the use of the electromagnetic analogy, they simply approach it in a characteristic way that emphasizes the points at which the analogy breaks down rather than the areas in which the comparison holds. They tend to employ analogy as a tool in a negative sense, rather than in a positive sense. Eddington, for instance, really expected that the gravitational field would propagate with the speed of light in the linearized approximation, but he was concerned to check for any differences from the familiar electromagnetic case. As he put it:

> This is the answer to the main question we have set before ourselves— whether the absolute gravitational influence, independent of co- ordinate systems, is propagated with the speed of light. This was perhaps a foregone conclusion. If we have an absolute gravitational disturbance which has entirely detached itself from the material sys- tem which originated it, only one thing can happen to it. It cannot stay at rest, since there is no absolute rest. It must travel; and since there exists only one speed which is independent of frames of refer- ence, it has no choice but to accept that speed. Our direct calculation verifies this prediction. (p. 272)[1]

It should be kept in mind that Eddington's argument that gravitational waves propagate with the speed of light applies, of course, only to the very weak gravitational waves that are described by the linearized approxima- tion. There is still no general proof that gravitational waves of arbitrary strength always propagate at this speed (in fact, there is a sense in which they do not always do so, since scattering off the curvature of spacetime can result in parts of a gravitational wave arriving "late," as a "tail" to the main wave, though this is not because it traveled more slowly, but rather because it backtracked for a while and traveled farther than the main wave). Later in the same paper, Eddington, having failed to construct a spherically sym- metric gravitational wave, devoted space to discussing "the reason for the breakdown of the analogy between sound waves [which can have spherical symmetry] and gravitational waves" (p. 276).

Although the coordinate choice Einstein made from the outset proved highly successful in motivating the analogy and produced a compelling description of gravitational waves, it also laid a trap for the unwary in the form of the writhing coordinates in which even flat space appears wavy. Eddington, with his innate skepticism, easily evaded the trap. He refused to take for granted that his waves would travel with a unique speed and quickly sniffed out the spurious variety, killing them off entirely with a geomet- ric argument that they represent flat space and showing that the Riemann tensor vanishes entirely for the spurious waves. Since the Riemann tensor describes the curvature of spacetime, when all its components are zero (van- ish), spacetime is flat. This is considered the surest, coordinate-independent way of determining that no gravitational waves are present, in the same way as in the case of the seaweed, where the simplest way to check there are no waves is to look at the surface of the water. Weyl, perhaps somewhat rushed in his treatment, fell straight into the trap, whereas Einstein, armed with his great intuition, carefully and laboriously navigated his way out of the snare.

Eddington thus stands in a skeptical tradition that begins with Abraham and was to be best personified by a physicist who regarded Eddington as his mentor in general relativity, Hermann Bondi. It was a tradition that exhibited much of what is best in science: a stubborn refusal to take anything for granted, an insightful ability to recognize the glib assumption in the most apparently obvious concept, and a careful eye for the limits of any result. It is a tradition that could be infuriating to some, although as with any scientific tradition, its best representatives, including Eddington and Bondi, were viewed with near universal respect, no matter from which quarter. In so far as gravitational waves are concerned, it was a tradition that was to have to fight hard to make its point over many decades, because increasingly, gravity waves were becoming one of those ideas that appear obvious and inevitable.

One gets the clearest impression of Eddington's way of thinking from his remarks in *The Mathematical Theory of Relativity* (Eddington 1923a), where he addressed the question of whether binary stars do or do not decay in their orbits due to gravitational radiation. In this case he was skeptical about skepticism, which, to judge from his remarks, may have been widespread at the time. This textbook was widely read and highly influential and was the principal way in which the particular flavor of Eddinton's approach to the relativity was handed down to a later generation of physicists:

No solution of Einstein's equations has yet been found for a field with two singularities or particles. The simplest case to be examined would be that of two equal particles revolving in circular orbits round their centre of mass. Apparently there should exist a statical solution with two equal singularities; but the conditions at infinity would differ from those adopted for a single particle since the axes corresponding to the static solution constitute what is called a rotating system. The solution has not been found, and it is even possible that no such statical solution exists. I do not think it has yet been proved that two bodies can revolve without radiation of energy by gravitational waves. In discussions of this radiation problem there is a tendency to beg the question; it is not sufficient to constrain the particles to revolve uniformly, then calculate the resulting gravitational waves, and verify that the radiation of gravitational energy across an infinite sphere is zero. That shows that a statical solution is not obviously inconsistent with itself, but does not demonstrate its possibility.

The problem of two bodies on Einstein's theory remains an outstanding challenge to mathematicians like the problem of three bodies on Newton's theory. (p. 95)

What is interesting about this is that Eddington implied that some relativists expected that binary stars would not emit gravitational waves, even though with their changing quadrupole moments, they appeared to be ideal sources. The motive for this belief seems to be that the existence of radiation damping in such systems would eliminate forever the possibility of finding a static, non-time-varying solution to the binary star problem, the kind of solution that lies at the heart of classical perturbation techniques in Newtonian celestial mechanics. In other words, it was taken for granted that a solution in which binary stars do not decay in their orbits due to gravitational wave emission, is possible; indeed, desirable. So at this early stage the skeptical position on radiation damping in binaries was almost the default one, although Eddington was too much the skeptic to accept any glib arguments for it. He implied that a number of people had already tried to do this by 1923.

Besides the technical reason for preferring that binary stars should not decay in their orbits, one might speculate that another motive encouraging belief in the existence of a static two-body solution was the same as that for Einstein's instinctive choice of a static cosmology. There is an ancient prejudice that the universe in general is static and unchanging. Perhaps astronomers had a bias in favor of stellar systems that did not decay, just as in the days of Laplace, and so they preferred to leave gravitational waves out of the picture when it came to such apparently natural producers of radiation. Keep in mind, also, that because the terms that drive the decay of stellar and planetary systems due to gravitational radiation only enter into the post-Newtonian equations of motion at the fifth order in the velocity of the system, it took some time before these terms were discovered. It was perhaps natural, in the early days, to assume that since only conservative terms appeared in the calculations that these were the only kind of terms there were. As Eddington stated, however, this merely meant that not enough work had been done to know which answer, static or decaying orbits, was correct.

In his 1922 paper Eddington derived, as Einstein had, the quadrupole formula. Indeed, he can be said to be the first person to write it down properly, because Einstein had made another, this time relatively insignificant, error in the calculation in his 1918 paper. Eddington's quadrupole formula differed from Einstein's by a factor of two. It is Eddington's which is correct, and having corrected Einstein in this regard, he pursued the matter a little further. Instead of looking at the wave generated by a source such as a spinning rod, what if we look at how the linearized equations govern the motion of the rod itself, taking into account the retarded nature of gravitational interaction? Is it possible to figure directly the energy lost by the source without dealing with the far distant waves at all? Eddington attacked this issue in a paper published in the *Philosophical Magazine* in 1923. It is probably the first paper in general relativity that investigated the question of radiation reaction in a system from the point of view of the problem of motion. Rather than focusing on the waves and appealing to an energy-balance argument, Eddington's approach in this paper was to calculate what effect the gravitational pull exerted by different parts of the rod on each other has on the rotational motion of the rod. By presuming in the calculation that it takes time for the gravitational attraction to propagate from one end of the rod to the other, it turns out, in a manner analogous to Laplace's calculation, that the rod's pulling at itself exerts a torque that gradually slows it down. The energy lost in this way can be presumed to have been radiated away as gravitational waves. The force exerted by one end of the spinning rod on the other is called the retarded force, because it appears to come from a previous (retarded in time) position of the far end of the rod and not from its current position:

> In this method we pay no attention to the waves receding into the distance, which by an analytical fiction are supposed to bear the energy [lost by the source system] away. (The waves are real enough, but their pseudo-energy-tensor is not regarded in the relativity theory as

real energy.) The loss of energy is considered from another aspect which has a certain historic interest. If gravitation is not propagated instantaneously the lag may introduce tangential components of the force opposing the motion of the system. This was the effect looked for by Laplace when he considered a finite velocity of propagation of gravitation in astronomical systems; from its non-appearance he deduced that the speed must be that the first order effect looked for by Laplace (proportional to v^2/c^2) is eliminated; but the result here found is actually the residual Laplace effect of the third order (proportional to v^6/c^6). The mutual [gravitational] attraction of the particles in the rod is, owing to the rotation, not exactly in the instantaneous line of the rod, and the resultant is a couple which slowly destroys the rotation. It will, however, be explained in due course that the present result does not solve Laplace's problem for an astronomical system, but is limited to cohesive systems. (Eddington 1923b, pp. 1112–1113)[2]

In this paper Eddington recovered the quadrupole formula again, agreeing with the result of his earlier paper. However, once again, the result was limited to cases where there was no chance of any experimentally detectable effect.

We have now encountered the main points of interest that, during later periods of controversy, drove the debate on the existence of gravitational waves and radiation damping in binary stars. It was straightforward to use the linearized approximation to describe weak gravitational waves, but it was not entirely clear how one could find a coordinate-invariant way to actually calculate how much energy they transport. Nor had any great thought been given to the question of how a source would be affected by the passage of such waves. How would they interact with matter? Of course one reason for early indifference to this question undoubtedly was the realization that gravitational waves were so exceptionally weak that there seemed no prospect of actually detecting them. Accordingly, the practical issues of their interaction with physical systems were deemphasized.

The issue Eddington had raised was the interaction of the wave and its own source, what is called back action, or radiation reaction. It is clear that the major interest here was whether binary stars might undergo decay in their orbits as a result of generating gravitational waves, what is known as radiation damping. Indeed it was precisely because of this question that the problem of the energy carried in the waves was addressed. Interestingly, there may have been an early prejudice in favor of the view that binary stars would not be damped in their motion due to radiation. This was also the central issue of the later controversies. The early calculations, involving the linearized approximation, were, as Eddington observed, actually inadequate to decide this question, because this approximation presumes a very weak gravitational field, in which the motion of the bodies is not constrained by their gravitational interaction. Quite the opposite is true in the case of binary stars, planetary systems, or other bound gravitational systems. Better approximations would be required to settle this question.

In addition, one had the choice of two different strategies. One was to calculate the total energy carried away by the waves from a source and then argue that the source must lose that much energy and calculate the consequences accordingly. As shown in the previous quote, Eddington had already pronounced himself a little skeptical of this approach, and later it was to be much criticized on several grounds. The most cogent of those objections, which we will examine in more detail later, was that beginning with an orbit that assumes the absence of gravitational waves (what is called a geodesic orbit) and then subtracting energy from the energy associated with such an orbit based on the waves a star in that orbit would produce presumes the result. How do we know that the actual orbit that the star follows, taking retarded gravitational effects into account, does not coexist with these "radiation" effects in such a way that the star does not lose energy over time at all? Perhaps the true orbit, as opposed to the fictitious geodesic orbit, does not radiate energy at all. In other words there is an obvious need to make such a calculation fully consistent. One way around this is to do a calculation of the type Eddington made for the case of the spinning rod, in which one actually works out the retarded self-interaction of the system. As we shall see in the case of binary stars, such calculations are very, very difficult to accomplish. It is this story which we will follow through the rest of the book.

5

Do Gravitational Waves Exist?

Albert Einstein had a number of confidants in physics with whom he could discuss his work in the most personal terms, although by 1936 one of his closest friends, Paul Ehrenfest, was dead, while others were in exile, as Einstein was. One of these particularly intimate friends was Max Born, who, like Einstein, had fled Germany upon the Nazis' rise to power. In a letter to his old friend, probably written in mid-1936, Einstein anticipated the arrival of a new collaborator whose previous work with Born had brought him to Einstein's attention; at the same time, he reported a surprising result:

> Next term we are going to have your temporary collaborator Infeld here in Princeton, and I am looking forward to discussions with him. Together with a young collaborator, I arrived at the interesting result that gravitational waves do not exist, though they had been assumed a certainty to the first approximation. This shows that the non-linear general relativistic field equations can tell us more or, rather, limit us more than we have believed up to now. (Born 1971, p. 125)

The young collaborator Einstein referred to was his assistant Nathan Rosen, with whom Einstein had worked for two years and had written two important papers. One was the famous Einstein-Podolsky-Rosen (EPR) paper, which formulated a paradox challenging the Copenhagen interpretation of quantum mechanics. In that paper Einstein had launched a final volley at the new quantum mechanics, one which would be argued about and chewed over by physicists for decades. The other was the paper that introduced the "Einstein-Rosen bridge," now known as a wormhole and associated, via the work of Kip Thorne and the novel *Contact* by Carl Sagan,

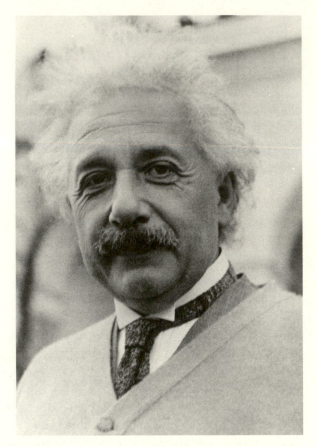

Figure 5.1. Albert Einstein, pictured during the 1930s. (Courtesy AIP Emilio Segré Visual Archives, Francis Simon Collection. Photo by Francis Simon)

with time travel. In the mid-thirties Einstein was living in Princeton, New Jersey, having permanently left Germany in 1933. At the recently founded Institute for Advanced Studies, he was provided with funds for a succession of young assistants who collaborated with him on his research, which mainly focused on his search for a unified field theory of gravitation and electromagnetism. Other subjects continued to hold his interest, however, and one of these was gravitational waves. Many of the assistants went on to distinguished careers in physics, and although the problems they chose to work on in later years were not always those that had occupied them during their time with Einstein, the topic of gravitational waves was an exception, as we shall see. In any case, towards the end of Rosen's time with Einstein they were interested in the question of what a plane gravitational wave looked like.

If we imagine a point source of gravitational waves, we see that the wave front propagating outwards will have a spherical shape centered on the source, but other shapes, or symmetries, are possible. For instance, when taking the source to be a line (one thinks of a piano string as an example), rather than a point (obviously these are all idealizations), we expect the wave front to be shaped like a cylinder as it radiates out. In the case of astronomy, of course, we expect that most sources will be so tiny, relative to their distance from us on the earth, that we can treat them as point sources, which produce a spherical wave front. Indeed, seen from earth, these sources are so far away that even the spherical shape of their wave front will not be apparent to us. It will appear as if the wave front is flat, because the radius of the sphere is so large that its curvature is very small. It will look like a plane wave. A true plane wave would have a source an infinite distance away, but the distant stars might as well be infinitely far away for our purposes. So if we are interested in detecting waves, especially in astronomy, we could most likely be interested in plane waves. Indeed, Einstein had chosen to describe plane waves in his early papers on gravitational waves, but those papers had dealt only with the approximate linearized equations of gravitation. Now he hoped to find an exact solution for plane gravitational waves that satisfied the full, nonlinear field equations of general relativity.

He submitted a paper by himself and Rosen to the *Physical Review* under the title "Do Gravitational Waves Exist?"[1] Einstein's answer to this question, as we can see from the letter to Born, was no. It is remarkable that at this stage in his career Einstein was prepared to believe that gravitational waves did not exist, all the more so because he had made them one of the first predictions of his theory of general relativity. We can get some idea of the surprising nature of this result from the autobiography of Leopold Infeld, Rosen's successor as Einstein's assistant. Infeld arrived in Princeton in the fall of 1936. He was a native of Poland, but being unable to find a professorship in his own country because of his Jewish origins, he had appealed directly to Einstein in a letter and was hired because Einstein was familiar with his successful collaboration with Born at Cambridge not long before. Although most of Einstein's assistants at this time were younger men, Infeld was nearly forty. His association with Einstein made his career and reputation. A popular book on science that they wrote together ensured his fame as the best-known of Einstein's assistants. But upon his arrival, Infeld could hardly bring himself to believe that Einstein thought that gravitational waves did not exist. As Infeld noted, by 1936 it was hard for any physicist to believe they did not, even though they had never been seen and even though most physicists expected that they never would be. In his first meeting with Einstein in Princeton, Einstein explained to him and to Tullio Levi-Civita, the notable Italian mathematician who was also present, why he no longer believed in the existence of gravitational waves. Levi-Civita was one of the founders of the field of mathematics (tensor calculus) that underlies general relativity and made it possible. In the earliest days of relativity he had criticized Einstein's use of a pseudo-tensor to describe gravitational field energy. Now, in the presence of two such great men, Infeld was amused as the "calm" Einstein and the "gesticulating" Levi-Civita stood at the blackboard and "talked in a language which they thought to be English." And despite his initial skepticism, Infeld soon convinced himself that Einstein was right by coming up with his own version of the proof, as physicists prefer to do (Infeld 1941, pp. 259–265).

But not everyone who was confronted with this new result was so easily convinced. Einstein had initially sent the paper to the *Physical Review* for publication, but it had been returned to him with a critical referee's report (EA 19-090), accompanied by the editor's mild request that he "would be glad to have your reaction to the various comments and criticisms the referee has made" (John T. Tate to Einstein, July 23, 1936, EA 19-088). Instead, Einstein wrote back in high dudgeon, withdrawing the paper and dismissing out of hand the referee's comments:

> Dear Sir,
>
> We (Mr. Rosen and I) had sent you our manuscript for *publication* and had not authorized you to show it to specialists before it is printed. I see no reason to address the—in any case erroneous—comments of your anonymous expert. On the basis of this incident I prefer to publish the paper elsewhere.
>
> Respectfully,
>
> P.S. Mr. Rosen, who has left for the Soviet Union, has authorized me to represent him in this matter. (Einstein to Tate, July 27, 1936, EA 19-086)[2]

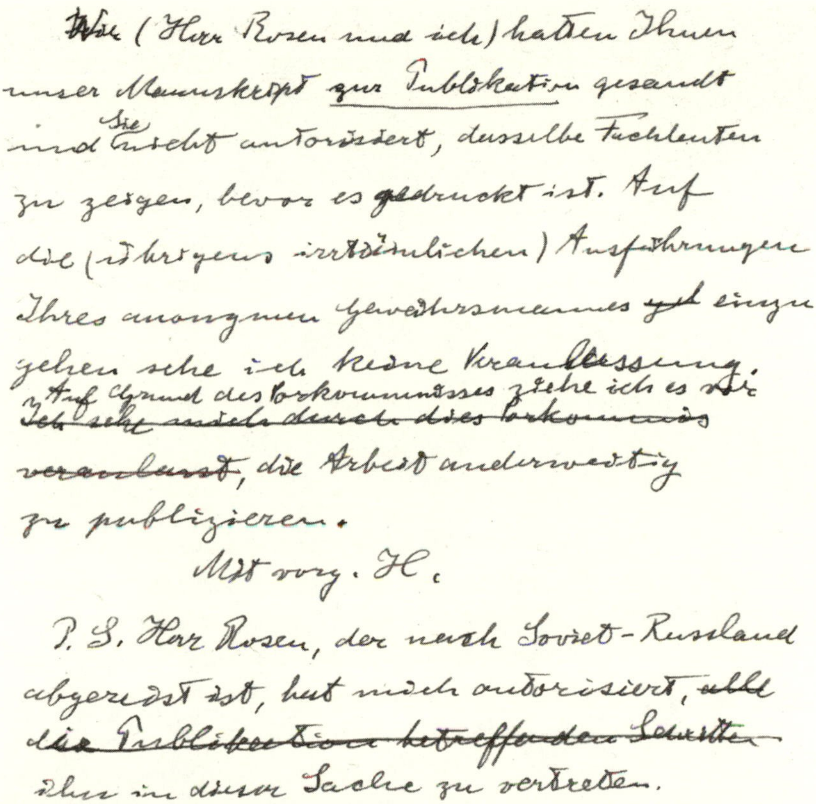

Wir (Herr Rosen und ich) hatten Ihnen unser Manuskript zur Publikation gesandt und Sie nicht autorisiert, dasselbe Fachleuten zu zeigen, bevor es gedruckt ist. Auf die (übrigens irrtümlichen) Ausführungen Ihres anonymen Gewährsmannes einzugehen sehe ich keine Veranlassung. Auf Grund des Vorkommnisses ziehe ich es vor ~~Ich sehe mich durch dies Vorkommnis~~ veranlasst, die Arbeit anderweitig zu publizieren.

Mit vorz. H.

P. S. Herr Rosen, der nach Soviet-Russland abgereist ist, hat mich autorisiert, ~~die Publikation betreffenden Schritte~~ ihn in dieser Sache zu vertreten.

Figure 5.2. Einstein's indignant letter to Tate, the editor of the *Physical Review*. (Courtesy Hebrew University of Jerusalem)

To this response, Tate, who had become editor as a relatively young man a decade previously and who was engaged in transforming the *Review* into the world's leading physics journal, replied that he regretted Einstein's decision to withdraw the paper but stated that he would not set aside the journal's review procedure. In particular, he "could not accept for publication in THE PHYSICAL REVIEW a paper which the author was unwilling I should show to our Editorial Board before publication" (Tate to Einstein, July 30, 1936, EA 19-089). Einstein's annoyance was such that he never published in the *Physical Review* again.[3]

In fairness to Einstein, the practice of using anonymous reviewers may have been unfamiliar to him. German journals in the early part of the century were considerably less fastidious than the *Physical Review* in what

Figure 5.3. John T. Tate, the editor of the *Physical Review* from 1926 to 1950, made the journal into the leading physics journal in the world. (Courtesy University of Minnesota Archives)

they published. Infeld claims that the German attitude, in contrast to that prevailing in Britain and America, was "better a wrong paper than no paper at all" (Infeld 1941, p. 190). In a letter to Einstein in March 1936, the relativist and fellow European exile, Cornelius Lanczos, remarked on "the rigorous criticism common for American journals," such as the *Physical Review* (translated and quoted in Havas 1993, p. 112). Historians who have studied the editorial policies of the *Annalen der Physik*, the leading German journal of the first decade of this century, in some detail note that "the rejection rate of the journal was remarkably low, no higher than five or ten percent" (Jungnickel and McCormmach 1986). They describe the editors' reluctance to reject papers from established physicists (p. 310). As the *Annalen* was the journal in which all of Einstein's early papers, from 1900 to 1905, were published, the "rigorous criticism" he was to experience very shortly after receiving Lanczos's letter must have come as something of a shock. At the same time, Einstein could be very frank and direct, albeit in private, in his criticism of other's work. From 1914 on, in his capacity as a member of the Prussian Academy of Sciences, he was regularly called upon to give opinions on the suitability of articles submitted to the academy's proceedings. At the Einstein Papers Project, which is publishing Einstein's collected papers, these little reviews by Einstein are known as the *Wertlos* documents, from the German word for "worthless," which frequently occurs there. Einstein himself, as a member of the academy, had his papers published without question or revision. Anything less must have seemed to him a tremendous slight.

The paper with Rosen was, however, subsequently accepted for publication by the *Journal of the Franklin Institute* in Philadelphia, a much smaller journal that Einstein had, however, published in previously. The paper appeared in the journal in early 1937 under a different title and with radically altered conclusions. That it had previously been accepted in its original form is indicated by a letter from Einstein to its editor on November 13, 1936, explaining why "fundamental" changes in the paper were required because the "consequences" of the equations derived in the paper had previously been incorrectly inferred (EA 20-217).

What had led Einstein to the conclusion that so surprised Infeld? Having set out to find an exact solution for plane gravitational waves, he and Rosen had found themselves unable to do so without introducing singularities into the components of the metric describing the wave. This was surely not at all what they had hoped for, but like good physicists confronted with the unexpected, they attempted to turn it to their advantage. In fact, they felt they could show that no regular periodic wavelike solutions to the equations were possible at all (Rosen 1937, 1955). Instead of a solution to the

Einstein equations, they had a nonexistence proof for solutions representing gravitational waves. This was a far more important and breathtaking
result! To understand a little about it one has to discuss the term *singularity*.

A singularity is a mathematical pathology that arises when one has a
point in a field or a function to which no definite arithmetical value can
be assigned. Usually the presence of a singularity is a sign that something
has gone wrong with the calculation, but it is all a matter of interpretation.
Some singularities are considered "real," in the sense that they point to the
presence of an actual physical phenomenon. A famous example of this is
the singularity that lies at the center of black holes, the point where the curvature of spacetime becomes infinite. In fact, even in Einstein's time, long
before black holes were properly understood, it was sometimes thought
appropriate to identify a singularity of the field with the presence of a
point mass, the theorist's artifact that has haunted physics since Newton.
But the field of a gravitational wave should not contain any masses, unless
they are the sources of the wave. Indeed, a plane gravitational wave has
only a notional source which is infinitely far away and not present in ordinary spacetime at all. Therefore Einstein and Rosen concluded that the
singularities they kept finding were pathological and a sign that there was
something amiss with the whole solution.

Today it is well known that one cannot construct a single coordinate
system to describe plane gravitational waves without encountering a singularity somewhere in spacetime. But it is also understood that such a
singularity is merely apparent and not real. It is what is known as a coordinate singularity. It tells you not that there is something wrong with your
solution, but only that there is something wrong with your coordinates,
which is quite a different matter. Take, as an example, the Earth's North
Pole as expressed in our usual coordinate system, that of latitude and longitude. What is the longitude of the North Pole? Every line of longitude
passes through the North Pole. It might as well be any number from zero to
three hundred and sixty degrees. Any answer is as good as another because
the system of numbering breaks down there. It is a singular point, but
there is nothing odd about the North Pole as a place. It is only the coordinate system that has failed, not our understanding of the shape of the
earth. A differently chosen system would have no difficulty describing the
location of the North Pole but might fail at another spot on the globe. It
all depends on the need of the moment. Einstein was one of the first to

understand the critical difference between coordinate and physical singularities, but in the thirties there was still no mathematical formalism for distinguishing between the two. It was something that had to be worked out by trial and, frequently, error. After the war the matter was settled by the work of mathematicians like the Frenchman André Lichnerowicz. The study of singularities in relativistic spacetime became much more rigorous, as the mathematicians say. In this case Einstein and Rosen were erring too much on the side of caution by treating a harmless coordinate effect as a real physical pathology in their solution. It simply did not occur to them that trying a one-size-fits-all approach in which one coordinate system was expected to cover the whole of spacetime was asking too much.

In August of 1936, the relativist Howard Percy Robertson returned to Princeton from a sabbatical year in Pasadena and in October (if we are to judge from Infeld's account, which tells us that Infeld arrived in Princeton on the day of a home football game) struck up a friendship with the newly arrived Infeld. Robertson was one of the most distinguished figures in the new field of cosmology during the period when the ground for what later became the big bang theory was being prepared, and when cosmology was considered about the only part of physics where general relativity had any relevance. Robertson was a colorful, jovial character who enjoyed cultivating enemies as much as he, in Infeld's words, "enjoyed spiteful gossip" about his colleagues. He probably did as much as Einstein to further Infeld's career, helping him to get his first real academic job. He told Infeld that he did not believe Einstein's result, and his skepticism was much less shakable than Infeld's. Certain that the result was incorrect, he went over Infeld's version of the argument with him, and they discovered an error (Infeld 1941, p. 241). When this was communicated to Einstein, he quickly concurred and made changes in proof to the paper, which was then with the publisher of the *Journal of the Franklin Institute* (Infeld 1941, p. 244; and Einstein to editor of the *Journal of the Franklin Institute*, November 13, 1936, EA 20-217).

Curiously, Infeld states that when he communicated to Einstein that he and Robertson had uncovered an error in his (Infeld's) version of the proof, Einstein replied that he had coincidentally and independently uncovered a (more subtle) error in his own proof the night before (Infeld 1941, p. 245)! Unfortunately Infeld does not give any details about these errors in his book. He does tell us that Einstein had only realized that his proof was incorrect and had still not managed to find the gravitational wave solution he had been looking for. But he had been closer than he thought all along, and it seems that it was here that Robertson made his key contribution, at least according to remarks made by Rosen in a later paper of 1955. (Although a footnote attached to the published version of Einstein and Rosen's paper acknowledges Robertson's help, it does not indicate its nature [p. 54].) It appears that Robertson observed that the singularity could be dealt with by a change of coordinates which revealed that Einstein and Rosen were dealing with a solution representing cylindrical waves. With this trick the worrisome singularity was relegated to the central axis of the spacetime, where one would expect to find the source of the cylindrical waves (the vibrating "string" that was emitting them). Associating singularities with a material source was relatively common and widely accepted, although Einstein himself and some others had often expressed serious reservations about the practice (Earman and Eisenstaedt 1999). But any port in a storm will do, and Einstein was happy to retitle his paper as "On Gravitational Waves" and present these cylindrical waves that he had stumbled upon unwittingly. The irony, of course, is that Einstein could have found this escape route months earlier simply by reading the referee's report which he had dismissed so hastily, because the "anonymous expert" had also observed that by casting the Einstein-Rosen metric (as we nowadays call this solution of the Einstein equations) in cylindrical coordinates, the apparent difficulty with the metric was removed, and it was seen to be describing cylindrical waves (Referee's report, Robertson Papers, California Institute of Technology, box 7.12; and EA 19-090, pp. 2, 3, 5).

But Einstein was not a man to waste time in embarrassment. Infeld relates the amusing detail that Einstein was scheduled to give a lecture in Princeton on his new "result," just one day after his discovery that his proof was no good. He had not yet spoken to Robertson to discover the way out of his difficulty, so he was obliged to lecture on the invalidity of his own proof, concluding by stating "If you ask me whether there are gravitational waves or not, I must answer that I do not know. But it is a highly interesting problem" (Infeld 1941, p. 246). Einstein rarely let personal pride interfere with his work. While working on the popular book which they wrote together, Infeld told Einstein that he took special care because he could not "forget that your name will appear on it":

> Einstein laughed his loud laugh and replied: "You don't need to be so careful about this. There are incorrect papers under my name too." (Infeld 1941, p. 316)

There remains one interesting, and only very recently uncovered, piece of evidence concerning how Einstein came to realize there was an error in his nonexistence proof. In the papers of the late Peter Bergmann (who became one of Einstein's assistants at about the same time as Infeld), there survived a very interesting partial draft of a paper in Einstein's handwriting titled "On Rotationally Symmetric Stationary Gravitational Fields." The paper is undated, but its opening section makes it easily dateable:

> Until recently it was justified to believe that to each solution of the linearly approximated gravitational equations there belonged an exact solution, which could be recovered from it by successive approximations. Recently I showed however, together with Mr. N. Rosen, that plane-wave solutions of the linearized equations do not correspond to any exact solutions. From this arises the suspicion that the diversity of the exact solutions is also in other cases more limited than it seems in accordance with the linearized equations. Thereby the problem of the integration of the gravitational equations acquires a deeper interest than seemed to be the case up to now.
>
> This consideration led me to undertake the following investigation of the rotationally symmetric special case. It is again shown here that the diversity of the solutions is limited far more than is to be expected from the linearized equations. (EA 80-974.0; translated from the German by the author)

This paper must have been written between June and October 1936. The version that survives is eleven pages long and ends abruptly at the beginning of the demonstration that a rotationally symmetric stationary solution of the exact Einstein equations does not exist. It is reasonable to conclude that Einstein stopped working on the paper when he realized that there was something wrong with his earlier paper with Rosen. One may speculate that it was the experience of working on what was to be the follow-up paper that gave birth to his doubts concerning the original. Perhaps he began to feel that the nonexistence of a stationary rotationally symmetric gravitational field was not a very likely result. Upon closer examination he may have felt that the proof was not convincing for reasons which also applied to the gravitational waves case. It is noteworthy that the rotational symmetry of the fields in the follow-up paper is similar to the cylindrical symmetry of the waves that eventually provided the counterexample to the earlier proof.

What is, in any case, particularly noteworthy about the introduction to this paper is Einstein's insistence that solving the exact equations of general relativity only becomes really interesting when it seems that it may impose stricter limits on the possible solutions the theory permits. This, of course, is precisely the sentiment he had expressed to Born in the letter quoted at the beginning of this chapter. At this time, general relativity was primarily of interest to Einstein in so far as it might provoke insights into the form of the more profound unified theory that must, in his view, eventually supplant it. In such an endeavor, with so many open possibilities before him, any clues as to the constraints that would enable him to find the right path could be invaluable. Indeed, for later theorists too, as we shall see, the possibility of finding radically new physics in relativity theory seemed often to rest with the hope of the unexpected result, such as the one that gravitational waves do not, after all, exist.

Although the solution for cylindrical gravitational waves now bears Einstein's and Rosen's names, it had in fact been previously published by the Austrian physicist Guido Beck in 1925, but his paper was completely unknown to relativists with the single exception of his student, Peter Havas, who entered the field only in the late fifties. In a 1926 paper by the English mathematicians Baldwin and Jeffery and in the referee's report on Einstein's paper, there was discussion of the fact that singularities in the metric coefficients are unavoidable when describing plane waves with infinite wave fronts. But although there is some distortion in the wave, "the field itself is flat" at infinity, as the referee noted (Robertson Papers, California Institute of Technology, box. 7.12; and EA 19-090, p. 9). Clearly the referee's familiarity with the literature exceeded Einstein's own, but then Einstein was notorious for being lax in this regard. Certainly the published Einstein-Rosen paper contains no direct reference to any other paper whatsoever and only a couple of other authors are even mentioned by name. When working with Infeld, Einstein responded to the latter's suggestion that he do a literature search for previous work by other scientists by laughing and saying, "Oh Yes. Do it by all means. Already I have sinned too often in this respect" (Infeld 1941, p. 277).

The question that naturally arises at this point is, who was the referee? The report is ten pages long and shows an excellent, if not perfect, familiarity with the literature on gravitational waves (the referee knew of Baldwin and Jeffery's 1926 paper but not Beck's of 1925). The copy forwarded to Einstein is typewritten and the spelling follows American practice (*behavior* rather than *behaviour*, "*neighborhood*" rather than "*neighbourhood*"). It is likely, therefore, that the author was an American with a strong interest in the theory of general relativity. Only a few people at this time fit that description, including Robert Oppenheimer and Richard Chase Tolman, both based in California. And suspicion naturally falls on Robertson himself; after all, he appeared to have the solution to the paper's flaws at his fingertips in the fall of 1936 when he spoke with Infeld. The report itself had been written while Robertson was on sabbatical at the California Institute of Technology, and therefore absent from Princeton. Caltech, located in Pasadena, near Los Angeles, was Robertson's graduate school and he was eventually to return there as a professor. Might he not have written the review there and only had the chance to approach Einstein in a more personal way upon his return to Princeton later in the year?

Unfortunately, when I first contacted the *Physical Review* in the mid-1990s, I was told they had no records from before 1938. Although Tate married his secretary, a situation that often leads to quite a thorough preservation of old papers, his personal papers apparently have not survived. But Robertson's own papers are preserved at Caltech, and when I first looked there as a graduate student, I found among them a letter to Tate, written on February 18, 1937, in which he said:

> You neglected to keep me informed on the paper submitted last summer by your most distinguished contributor. But I shall nevertheless let you in on the subsequent history. It was sent (without even the correction of one or two numerical slips pointed out by your referee) to another journal, and when it came back in galley proofs was completely revised because I had been able to convince him in the meantime that it proved the opposite of what he thought.

You might be interested in looking up an article in the Journal of the Franklin Institute, January 1937, p. 43, and comparing the conclusions reached with your referee's criticisms. (California Institute of Technology, Robertson Papers, folder 7.13)

So it turns out that Robertson was the referee. Finding that Einstein had completely ignored his written critique, he took the opportunity of their collegial closeness at Princeton to pass on his advice in a less confrontational fashion than was possible as the paper's anonymous referee. It is apparent that Robertson did not tell Einstein that he had refereed the paper; he did not even tell Infeld, with whom he became rather friendly. Once one knows about Robertson's previous close reading of Einstein's paper, Infeld's description of their discussion evokes a smile:

> The next day I met Robertson in Fine Hall [home of the Princeton University mathematics department] and told him: "I am convinced now that gravitational waves do not exist. I believe I am able to show it in a very brief way." Robsertson was still skeptical: "I don't believe you," and he suggested a more detailed discussion. He took the two pages on which I had written my proof [Infeld's own demonstration of Einstein and Rosen's result] and read it through. "The idea is O.K. There must be some trivial mistake in your calculations." He began quickly and efficiently to check all the steps of my argument, even the most simple ones, comparing the results on the blackboard with those in my notes. The beginning checked beautifully. I marveled at the quickness and sureness with which Robsertson performed all the computations. Then, near the end, there was a small discrepancy. (Infeld 1980, p. 267)

The 1937 letter from Robertson to Tate was the only firm evidence I could find that he was indeed the referee, until the Einstein Year, 2005, when a colleague put me in touch with the current editor of the *Physical Review*, Martin Blume, who had a very exciting discovery to report. It turned out that the logbook used by the *Review* in the thirties and forties had been uncovered, and thanks to the kindness of Dr. Blume in revealing the contents of the relevant line from the logbook, this find enabled us to confirm that Robertson was indeed the referee. The logbook shows the dates of the various stages in a paper's publication process after submission

NAME	DATE IN	REFEREE	DATE IN	TO AUTHOR	TO N.Y.	ISSUE	RE-JECTED
1936							
Glasgow	5/24	*Turstin 6/14*	6/18				6/13
Einstein & Rosen	6/1	*Robertson 7/6*	7/17	7/23			
Q........	-6-5				4/4	MAY 15, 1936	
Grumor & Serier	3/28		4/16	4/8	4/17/36	JUNE 1, 1936	

Figure 5.4. The logbook of the *Physical Review*, dating from the 1930s. The line shown gives Robertson as the name of the referee of Einstein and Rosen's paper. (Courtesy Martin Blume and the *Physical Review*)

to the *Physical Review*, along with the name of the referee, which in this case is Robertson's. Although the identity of the journal's referees is normally a highly confidential matter, in this case Dr. Blume was willing to confirm my guess because all of the people involved had passed away. The article was received by the journal on June 1. After a delay of more than a month, it was sent to Robertson on July 6 and returned to Tate on July 17. Five days later (July 23) it was sent back to Einstein, a date that is confirmed by Tate's letter to Einstein cited above.

Inspired by this discovery, I returned to the Robertson archives to check on his movements that summer. Though they were colleagues at Princeton, it seemed of interest to confirm that Robertson and Einstein had not been in personal contact throughout the period of the paper's gestation (if we assume that was the first half of 1936). To my surprise further material had been added to the archive since my previous visit, ten years earlier, and the collection had been restructured. Sitting in the middle of the Tate correspondence that I had looked through before was most of the actual correspondence between Robertson and Tate concerning the paper! It confirmed that Robertson had received the paper in Moscow, Idaho, where he was visiting after leaving Caltech and before returning to Princeton. Here is what Robertson had to say in his reply, dated July 14, to the still-missing original of Tate's letter enclosed with the paper:

Dear Tate:

Well, this *is* a job! If Einstein and Rosen can establish their case, this would constitute a most important criticism of the general theory of relativity. But I have gone over the whole thing with a fine-tooth comb (mainly for the sake of my own soul!), and can't for the life of me see that they have established it. It has long been known that there *are* difficulties in attempting to treat *infinite plane* gravitational disturbances in general relativity—even in the classical theory the potential acts up at infinity in such cases—and as far as I can see the additional, much more serious, objections of Einstein and Rosen do not exist. I can only recommend that you submit my criticisms to them for their consideration, and with this in mind I have written up in duplicate a series of "Comments" which you can, if you are so minded, send them. The alternative would be to publish it as it stands, taking account only of Comments (a) and (b) which deal with typographical errors of a minor sort. Such a paper would be certain to give rise to a lot of work in this field of gravitational waves, which might be a good thing—provided they didn't flood you out of house and home, as in the case of the Page excitement! If you decide to refer it back to the authors with my comments, please send them the copy written on white paper, keeping the yellow paper copy for your files.

Accompanying this letter in the Robertson archive is a one-page cover to the report, which was not sent to Einstein, and the yellow paper copy of the ten pages that were shown to the authors and are preserved in the Einstein Archive (for a transcription of the entire report, with its cover page, see appendix A).

In his reply, dated July 23, Tate thanked Robertson for his "careful reading of the paper" and rewarded his diligent referee in the usual manner, by sending him another tricky assignment.

In his letter to Einstein, Tate had carefully avoided stating that anonymous review by the editorial board or others was a necessary step in the acceptance of a paper by the journal. In fact the logbook (again, thanks to Dr. Blume for this information) shows that neither of the two previous papers by Einstein and Rosen (including the one with Podolsky) had been sent to a referee. At least in both cases, the field for the referee's name is left blank, and in the case of the EPR paper the paper was sent to New York for publication the day after its receipt from the authors. Therefore it is likely that the gravitational wave paper was genuinely Einstein's first encounter with the anonymous peer review system as practiced in American journals at that time (which was, in any case, still only emerging as the system we know today and was not yet the general rule it has become).

Figure 5.5. Howard Percy Robertson, the author of the referee's report that annoyed Einstein but that proved correct. (Courtesy AIP Emilio Segré Visual Archives, *Physics Today* Collection)

It is interesting that Tate chose to have this paper refereed. After all, the two previous submissions were certainly controversial. The EPR paper is arguably the most controversial paper Einstein ever published. The Einstein-Rosen bridge paper was part of an ongoing controversy with Ludwig Silberstein (Havas 1993). Einstein and Rosen's letter to the *Physical Review* in 1935 was part of this same debate. But Tate was willing to publish both of these papers relying on his own opinion. A paper, however, that claimed to prove that gravitational waves did not exist apparently sounded alarms with him. It is commonplace nowadays to imagine that most physicists cared little for and knew less of general relativity in this period, but apparently gravitational waves were already such a well-accepted prediction of the theory, in spite of the absence of experimental support, that such a surprising result required some scrutiny. It is noteworthy that more than a month seems to have elapsed between receipt of the paper and its referral to Robertson. This certainly suggests hesitation on Tate's part and may even be evidence of an initial round of editorial discussion.

That Tate often relied on his own instincts as editor is suggested by an anecdote related by the future Nobel laureate John van Vleck that in the early days of matrix mechanics the refereeing of one 1926 paper "consisted of Tate's showing him [Van Vleck] the manuscript in the Physical Review office" Nier and Van Vleck 1975). In general Tate did not like to slow the publication of important work. It certainly seems that he had good instincts regarding Einstein's contributions. He published two of his better-known papers expeditiously and, by choosing Robertson as referee for the third, saved him from what would have been a very public embarrassment, since the relatively innocuous paper that was published still attracted newspaper attention. It is unlikely that Einstein would have cared much one way or the other. Tate, however, paid the price of never again receiving a submission from "his most distinguished contributor."

Rosen was in the Soviet Union throughout this whole episode. Einstein had helped him to secure a position in Kiev, even writing to Vyacheslav Molotov on his behalf in early July, from which letter we learn that Rosen did not leave the United States until later that month. He continued to correspond with Einstein, his letters full of praise for the Soviet system, until his abrupt announcement that he was returning home to the United States (Rosen to Einstein, July 31, 1938, EA 20-227). Although in his letter Rosen gave as the reason for his return the feeling of being "dissatisfied with my own work and ability," one imagines this sentiment was written largely for the benefit of the censors and that his decision arose from a prudent desire to leave the Soviet Union, then enduring the worst phase of the Stalinist purges against foreign intellectuals.

His letters to Einstein reveal that not only had Einstein discovered the error in the paper simultaneously with and independently of Infeld, but so had Rosen! One can imagine his surprise upon learning that the paper had appeared in a different journal than the one to which he and Einstein had submitted it. Full of concern he wrote to Einstein in February 1937:

Dear Professor,
    I have just received from a friend a newspaper clipping in which it is stated that our paper, revised by you, has appeared in the *Journal of the Franklin Institute*. This journal has not reached us yet and newspaper

articles are of course not very reliable, but I get the impression that you did not receive my letter which I sent some time ago—which perhaps accounts for your not answering. At the time, among other things, I pointed out that our conclusions had been incorrect. (Rosen to Einstein, February 26, 1937, EA 20-218).

It should be noted in passing how rare it is for physicists to read news of their collaborator's resubmittal of their esoteric paper to a less prestigious journal from a newspaper report! When Rosen did read the article, he was not terribly happy with the turn which events had taken. The revised paper seemed to him to dodge the central question: "In the published paper the error [in our reasoning] is avoided—but at the cost of avoiding the problem. The question is raised: are there plane waves? and the answer is given: yes there are cylindrical waves." Accordingly, the same year (1937) he published a paper in a Soviet journal, carrying through what is presumably the chief argument of the original version of the Einstein-Rosen paper, in order to show that plane gravitational waves did not exist because of unavoidable singularities in the metric. After the war it was shown by Hermann Bondi, Felix Pirani, and Ivor Robinson that Rosen's singularities were not real and that plane gravitational waves do exist in relativity theory.

Did the retraction of his nonexistence proof mean that Einstein was cured of his skepticism? Perhaps not, or at any rate, not completely. Even if gravitational waves exist, does it necessarily follow that binary star systems must radiate them and thereby lose energy and decay (however slowly) in their orbits? Even in the published version of the paper there is a long section devoted to this question. Not necessarily, was Einstein's answer. The waves emitted by a source need not damp the source's motion, Einstein said, if one supposes that any outbound radiant energy is matched by a second system of incoming waves, impinging on the source. In technical terms, the use of half-advanced-plus-half-retarded potentials prevents damping of the source system even if the waves are emitted. In the authors' words, "this leads to an undamped mechanical process which is embedded in a system of standing waves" (Einstein and Rosen 1937, p. 48).

What does this mean? A potential refers to an aspect of the field generated by a source, whether a charged particle in electromagnetism or a massive body in the case of gravity. A retarded potential is one whose effect propagates at a finite speed. It is called retarded because the potential felt by another body is the potential produced by the source body at a previous "retarded" time, which depends on the time taken by the field effect to travel the distance between the two bodies. This is simply because of the time taken for the field to transmit the information about the source's current position. There is always a lag in the information available. An advanced potential is one in which the receiving body senses the field produced by the source at a future point in its motion. It is just the same concept in reverse. Obviously advanced potentials violate our sense of causality, the order in which things happen. Cause should precede effect, not the other way around. It is a feature of field theories that these advanced solutions are just as valid as the more natural retarded ones. Such theories are called time symmetric.

Einstein and Rosen refer at this point to the work of Walter Ritz, a Swiss contemporary and friend of Einstein's. Ritz had criticized electrodynamic theory on the grounds that advanced potentials run so counter to our experience of reality that they should not even be admitted as solutions of the equations of electrodynamics (Ritz 1908). He thought that the "true" equations of electrodynamics would not allow such solutions. An example of the oddities that would arise if such potentials existed in nature is that waves would appear to work backwards in time. Instead of starting at a source and rippling out in all directions towards infinity, they would start at infinity and ripple in from all directions towards what would now be the receiver. Such waves would bring energy with them, rather than carry it away.

What happens if, as Einstein proposed, you arrange matters so that the solution to your field equations is a mixture of equal parts of advanced and retarded potentials (half-advanced-plus-half-retarded)? Then the energy loss to the outgoing waves and the energy gain from the incoming ones balance each other and you have what Ritz would have objected to as a perpetual-motion machine. Your binary star would never decay, revolving endlessly in its orbit as the ancients imagined. Indeed, the waves themselves would appear to be standing still, balancing each other precisely, so that one can also refer to this as a standing-wave potential. Doesn't this also violate our experience of nature? Ritz asked. The arrow of time always points in only one direction, yet the field equations appear to insist that both directions are equally valid. Yet Einstein talked of this possibility as if he regarded it as a plausible scenario.[4]

The Dutch physicist Hugo Tetrode, also an acquaintance of Einstein, discussed the standing-wave potential in a paper of 1922, and Einstein and Rosen mention his name along with that of Ritz. At the time, this solution to the classical wave equations seemed a possible explanation for the failure of orbiting atomic electrons to radiate continuously, which would cause atoms to quickly collapse as the orbits decayed. Furthermore, Tetrode pointed out, in the quantum regime, the emission and absorption of radiation seemed to each depend on the other, rather than emission being required for absorption, but not the reverse. This suggested to him that the classical aversion to making absorption a requirement for emission should be discarded: "The Sun would not shine if it were alone in the universe" (Tetrode 1922, p. 323). Tetrode's theory can be seen as an action-at-a-distance theory, because the net effect of combining the two potentials is that one starts off by assuming that distant bodies feel the source's field with no time lag.

Although Ritz regarded the retarded potential as the only legitimate one, nevertheless the type of emission theory he advocated is often linked with action-at-a-distance theories, because it proposes that light emitted by a source partakes of the motion of the source. This means that instead of the expanding spherical wave front of the light being centered on the retarded position of the source, it is centered on the source's current position (assuming the source is unaccelerated), because the light is moving along with the source. Thus the light appears to be coming from the source's current position, as is characteristic of action-at-a-distance theories (see figure 5.6). It is possible that it is for this reason that Einstein here lumped Ritz and Tetrode together, though at first glance their views appear divergent. It is known that Einstein had given a great deal of thought to emission theories during his development of the special theory of relativity prior to 1905.

In the years immediately after Einstein wrote this paper, John Wheeler, then a young field theorist also at Princeton and later to become the most influential of all relativity theorists after Einstein himself, also, but completely coincidentally, tried to develop a coherent action-at-a-distance theory of electromagnetism. With his student, Richard Feynman, the future Nobel laureate, he began with the apparently acausal, half-advanced-plus-half-retarded potential describing a source with a great number of possible receivers. They found that the mere presence of many receiving bodies in the universe forces the equations to break down into a retarded-potential-only solution. What happens is that the receivers respond to the signal by emitting their own fields, which act back on the source. This back action breaks down the symmetry of the potential by canceling out the advanced half of it so that only the retarded part survives. Thus, time recovers its arrow. Just as Tetrode had put it, the ability to shine depends on something being out there to see it. Even more impressively, what began as a purely time-symmetric calculation turned out to exhibit exactly the behavior we experience in nature, thus apparently answering Ritz's objection.

Wheeler worked at Princeton for most of his career, and in his autobiography he tells us that when he and Feynman described their results for the

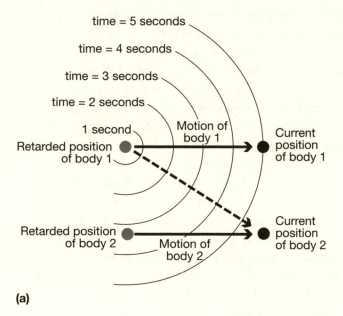

(a)

Figure 5.6. Ritz's theory as action-at-a-distance theory. (a) Concentric circles illustrate wavefront of light emitted by body 1 at intervals of 1 second. Arrows indicate path of light rays observed by a body moving parallel to body 1 at intervals of 1 second. If light does not partake of its source's motion, then the expanding wavefront of light remains always centered on the position at which it was emitted, and observing bodies perceive the light as originating from that retarded (past) position.

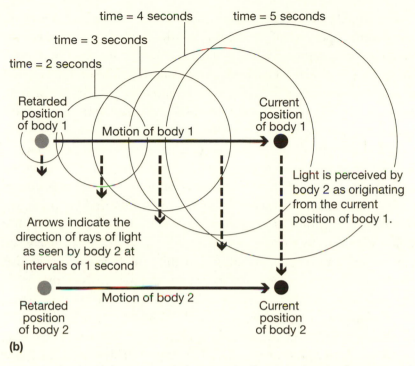

Figure 5.6. (b) Each successively larger circle illustrates the wavefront emitted by body 1 at intervals of 1 second. Arrows indicate the direction of rays of light as seen by body 2 at intervals of 1 second. If light partakes of system's motion, as in an emission theory, then the center of the expanding wavefront moves along with the motion of body 1.

elderly Einstein in the late forties, the old physicist responded, "I have always believed that electrodynamics is completely symmetric between events running forward and events running backward in time. There is nothing fundamental in the laws that makes things run in only one direction. The one-way flow of events that is observed is of statistical origin. It comes about because of the large number of particles in the universe that can interact with one another" (Wheeler 1998, p. 167). Einstein had put this opinion in writing in a brief statement with Ritz in 1908 (an agreement to disagree between the two friends), saying that the field equations were completely indifferent to the direction of time, but that only an ensemble of interacting bodies, as in thermodynamics, could give rise to an arrow of time.

How should we read the passage cited earlier from page 48 of Einstein
and Rosen's paper on half-advanced-plus-half-retarded potentials? If gravitational waves exist after all, why should we doubt that they are emitted
by binary star systems? The flavor of the claim is reminiscent of the kind
of attitude summarized in Eddington's book, quoted in chapter 4, and we
see here once again the early bias in favor of the idea that there should be,
as in Newtonian theory, in which gravitational waves do not exist, a static,
eternal solution for the motion of two stars about each other. We can surely
regard this passage as originating with Einstein, since the revised version
of the paper was written long after Rosen's departure for the Soviet Union.
Even if this passage is a survival from the original draft of the paper, the
references to Ritz and Tetrode certainly point to Einstein's authorship. It is
difficult for us today to imagine that Einstein was arguing that binary systems are undamped by gravitational wave emission, when it is obvious that
the electromagnetic analogy suggests that retarded potentials are the correct
physical choice, no matter by what argument you arrive at this conclusion.
But we have to concede that those close to Einstein, both Infeld (1960)
and Rosen (1979), seem to have taken this to be his argument. According
to one of his later students, Andrzej Trautman, Infeld often claimed that
Einstein supported his (Infeld's) own contention that binary stars did not
radiate gravitational waves. In 1979 Nathan Rosen published a paper with
the title, "Does Gravitational Radiation Exist?" which is evocative of the
never published version of his paper with Einstein. In it he argued that
the Wheeler-Feynman result, which works for electromagnetism, does not
work for gravitational fields. His calculation tried to demonstrate that the
gravitational field was too weak (it is notoriously far weaker than the electromagnetic field) for the absorbers of radiation to act back on the source
effectively and break down the half-advanced-plus-half-retarded field into a
retarded only field. Thus, in the gravitational case, the large number of bodies in the universe are too loosely coupled to each other by the gravitational
field to break down the time symmetry inherent in the equations. The
result would be, just as Infeld claimed almost to his dying day, that binary
stars would not radiate gravitational waves (and neither would anything
else, Rosen asserted, essentially because nothing can absorb gravitational
waves effectively.) However, there would still be gravitational wave fields,
the standing waves of the solution, so that detecting gravitational waves
might be possible. This was the most elegant argument ever made for this

position, but alas for Rosen, ever the gadfly of physics with his provoking and daringly argued proposals, it came far too late to have any real impact on the debate. It should be added that the Wheeler-Feynman absorber theory itself, although widely admired and much discussed, has never become a commonplace tool of the physics of radiation, either electromagnetic or gravitational, so it is perhaps not surprising that Rosen's argument received little notice.

Speculating on one possible motivation for this standing-wave passage, I refer back to Einstein's paper of 1916, his first paper on gravitational waves. Having derived his incorrect formula for energy loss by a source of gravitational waves, Einstein remarked in closing:

> Nevertheless, due to the inneratomic movement of electrons, atoms would have to radiate not only electromagnetic but also gravitational energy. As this is hardly true in nature, it appears, that quantum theory would have to modify not only MAXWELLIAN electrodynamics, but also the new theory of gravitation.

Although Einstein was referring here to his incorrect "monopole" formula for gravitational radiation, it is clear that atomic electrons would also, in general, radiate under the correct quadrupole formula of his 1918 paper, where he made a similar comment. Note also Einstein's strongly analogic style of reasoning. Although it was a mystery why atoms did not decay as a result of electromagnetic radiation damping, any gravitational radiation damping that they would experience would be infinitesimal by comparison. When Einstein said that it is "hardly true in nature," he was not worried that this fact had not been checked experimentally! It was true in the electromagnetic case and it follows that it must hold in the gravitational case also.

At this time Einstein naturally expected, as most later physicists were to do, that a quantum theory of gravity would eventually supersede general relativity theory. By 1936 his attitude to the new quantum theory which had emerged in the meantime had, as is well known, become entirely unsympathetic. He now hoped he could go in the opposite direction and develop general relativity into a unified field theory that would make the new quantum theory irrelevant, or at least explain it away. Perhaps, looking back to the one area of his theory that he had identified as in need of quantum correction, he thought to revive an old classical explanation of the undamped motion of orbital electrons, which was that their motion was somehow governed by a standing wave potential. Whereas the Einstein of 1916 was looking for areas in which progress towards a new quantum theory could be made, the disillusioned Einstein of 1936 was looking backwards to possible classical escape routes from what he saw as the dilemma of the new quantum mechanics.

Certainly interest in quantum gravity, which helped generate renewed interest in general relativity in the 1950s, also generated work on gravitational waves. For instance, Pirani (decidedly a nonskeptic) stated:

> The primary motivation for the study of [gravitational radiation] theory is to prepare for quantization of the gravitational field. (Trautman, Pirani, and Bondi 1965, p. 368).

Ironically, even though gravitational waves never yielded a major breakthrough in quantum gravity research, this renewed interest breathed new life into the classical theory of general relativity, a result that no doubt would have pleased its creator greatly.

At this point Einstein's personal quest for gravitational waves came to an end. He left a rich but ambiguous legacy. On the one hand, his was the first description of the phenomenon, and the formula that describes the flux of gravitational wave energy emitted by a source will be forever associated with his name. On the other hand, his views from the 1930s, however difficult it is to understand the nuances today, seem to have greatly inclined his assistants Infeld and Rosen towards skepticism. They were especially doubtful about whether gravitational waves would carry energy away from binary stars. Whether Einstein shared their concerns in the way they later formulated them is hard to know, because most of the subsequent debate took place after his death. But Infeld and Rosen, in their own work and through their students, had a great impact on the future development of gravitational wave theory, and their experience with Einstein can be credited with putting the questions in their minds that they later contributed to answering. But in 1937, this was all well in the future. The battleground had been laid out, but the forces that would fight on it had hardly yet begun to assemble.

6

Gravitational Waves and the
Renaissance of General Relativity

Einstein and Rosen had not been the first to find an exact solution of the equations of general relativity representing gravitational waves. Others had done so much earlier, in the mid-twenties. That their work was not known to Einstein is partly explained by his notorious disregard for the literature, which became increasingly pronounced in his later years, as Infeld discovered. However, the earlier papers on exact solutions of gravitational waves also were forgotten by relativists after the war for the simple reason that there was an interregnum in research in general relativity theory that lasted for a period of thirty years, from the mid-twenties until the mid-fifties. In the first ten years after Einstein's completion of general relativity, there were a number of researchers, mostly in Germany or neighboring countries (in particular Switzerland and Holland), who worked on the theory, and it enjoyed a period of rapid development. But with the advent of quantum mechanics in the mid-twenties, it entered a period of near total eclipse in physics. The problem was partly the extreme paucity of experiments whose results required general relativity for their correct interpretation (Eisenstaedt 1986a, 1986b, 1993, and 2006 for an account of this period in the history of general relativity). While quantum mechanics addressed a vast and growing field of experimentation, from established fields like atomic theory to new and exciting ones such as nuclear and particle physics, general relativity was relevant to the interesting but then somewhat limited field of cosmology and to almost nothing else (although the field of observational cosmology had begun to open up in the first decades of the twentieth century, it did not really take off until after the Second World War).

The impossibility of doing experimental work in the field made it extremely unpromising also for theoreticians, who preferred to do calculations that stood a reasonable chance of being vindicated by experiment in their lifetimes. It is hardly surprising that most of the people active in relativity theory (meaning the general theory unless otherwise stated) in this thirty-year period between 1925 and 1955 were mathematicians. In mathematics, vindication does not come from experiment but from providing rigorous proofs of theorems. Mathematicians recognized in relativity, from its inception, a physical theory that was well suited to their way of working. Its beautiful formal structure, while forbidding to many physicists, even theory-minded ones, was in many respects a straightforward product of nineteenth-century mathematics. Much important work was done during this period in mid-century when relativity found refuge in the mathematics departments of a few universities, but one thing that was not much pursued was the subject of the emission of gravitational waves from physical systems such as binary stars. The sort of approximate methods required to tackle such a task did not facilitate the statement of rigorously provable theorems and therefore did not prove attractive to mathematicians or to mathematics-minded physicists. Approximate calculations, by their very nature, may be wrong. If they are, usually the only way to find out is by experiment or observation, or else one can perform more approximate calculations to check them. If the experiments cannot be performed, then a great incentive for making the calculation is missing.

The fact that relativity was, for a considerable time, dominated by scientists with a strong background in mathematics was to have an important influence on the future controversies surrounding gravitational wave theory. When the theory began to move back into the mainstream of physics, there was an inevitable culture clash. The mathematical background of many relativists bred a long-standing attitude among physicists that general relativity was really a branch of mathematics with little relevance to physics. Even relativists who were trained as recently as myself have heard comments to the effect that general relativity is not really part of physics. On the other hand, it might also be argued that mainstream physics has moved in the direction of relativity. Nowadays some thriving areas of research in theoretical physics, such as string theory, have even less contact with experiment, and even greater relevance to mathematics, than used to be the case with relativity in the mid–twentieth century.

Relativists who played a leading role in the revival of general relativity later recalled the unique character of the field in the thirties and forties. André Lichnerowicz began working in relativity in 1937, at which time

> the relativity community had a strange mix. There was a small group of specialized physicists, Einstein's friends or students, such as W. Pauli, L. Infeld, B. Hoffman, and V. Fock (USSR); and a small group of specialized astronomers, such as G. Lemaitre; and a small group of mathematicians, such as T. Levi-Civita, T. de Donder, and G. Darmois.... The physics community of the time, passionately involved with quantum mechanics, considered the relativists to be marginal.... This state of affairs lasted until 1955 and the Berne conference presided over by Pauli. (Lichernowicz 1993, p. 103)

The conference at Berne conventionally marks the turning point in the fortunes of general relativity as a theory (Eisenstaedt 1986a, 1986b). It was held as a jubilee in honor of the fiftieth anniversary of the discovery of special relativity as well as the fortieth anniversary of general relativity itself. By coincidence it took place in the year of Einstein's death. Despite its commemorative aspect, in practice it was attended mostly by physicists interested in working in relativity theory and who were excited, perhaps even surprised, to meet such a large number of fellow enthusiasts.

The organizer of the Berne conference, André Mercier, hesitated before organizing it, because of negative attitudes on the part of other physicists who felt that the

> theory of general relativity had...become so much of a game consisting of playing with problems of a mathematical nature, that an acceptable connection with "real" physical problems had been lost and practically most of the scientists working in that field were not or were no longer attached to physical laboratories, but to mathematical institutions. (Mercier 1993, p. 110)

Thus, up until the late fifties, general relativity retained a marginal position within the physics community, with a small research community scattered over many countries (the United States, France, Germany, Britain, Ireland, the Soviet Union, Poland, and Israel, to name the most important in no particular order). The result was that communication between relativists was weak or nonexistent and exacerbated problems stemming from the lack of an institutional basis for the subject. There were no dedicated journals and no conferences or meetings devoted to general relativity that could provide forums for researchers to become aware of each other's work.

For this reason much of the work done on gravitational waves (and, indeed, on the problem of motion, relevant to binary star systems) between the wars, was simply forgotten by the time relativity began to undergo its revival in the wake of the Berne conference, but brief mention of the key work done in the twenties and thirties is still in order. The first publication of the exact solution for plane gravitational waves sought by Einstein and Rosen is due to O. R. Baldwin and George B. Jeffery. Their paper, published in 1926, is interesting in that it devotes much space to demonstrating that the spurious longitudinal-longitudinal and transverse-longitudinal waves of Weyl's textbook are really flat space even in the exact theory (as opposed to the linearized approximate theory) and that the plane transverse waves, as shown by Eddington, represent real propagating space-time curvature. Otherwise the paper is focused more upon the question of the propagation of electromagnetic waves (light) through curved space-time, something that Eddington had also stressed and a topic of obvious relevance to astronomy and other areas of physics, one which continued to be widely studied after the Second World War. The solution known to relativists as the Einstein-Rosen solution for cylindrical waves actually had been published earlier by a young Austrian doctoral student named Guido Beck. Beck was one of many scientists who turned from work in relativity to quantum theory or electrodynamics after completing their doctorates. Such changes were typical of the time. Quantum mechanics had just been born, and by comparison general relativity appeared a stone-cold research area. Beck, like many other European physicists, subsequently endured the dislocation of exile, being forced to flee his native Austria when the Nazis came to power. Interned in France during the war, he eventually made his way to South America. As it turned out, he would nevertheless influence the field of relativity in a very indirect sense, via his student, Peter Havas, who worked with him in France and who years later turned to relativity theory

Figure 6.1. Lev Davidovich Landau (known as Dau), one of the most brilliant physicists of the mid-twentieth century, with one of his students, and coauthor, Evgeny Lifshitz. (Courtesy AIP Emilio Segré Visual Archives, *Physics Today* Collection)

and, as we shall see, played a leading role in the controversies surrounding the radiation problem.

The most influential contribution to the study of gravitational waves dating to the interwar period was by the celebrated Russian duo of Lev Landau and Evgeny Lifshitz. In their famous textbook on classical field theory, they directly addressed the question of emission of gravitational waves from a binary star system and concluded that such a system would radiate energy in accordance with Einstein's quadrupole formula.[1] Their book was published in Russian in 1941, with a second edition in 1948, still a period when general relativity was in near total eclipse as far as most physicists were concerned. The book first appeared in English in 1951, translated from the second Russian edition. Because the influence of Landau in the Soviet Union was so great (as was that of others with similar views on the existence of gravitational waves, such as Vladimir Fock, of whom more later), there was little skepticism regarding gravitational waves there, so I will concentrate on the English edition and the reaction to it. (For a general discussion of these famous textbooks, see Hall 2005.)

While influential, the textbook's treatment of gravitational waves was also controversial. Indeed, attitudes towards it appear to have been so divergent as to make a striking example of the differences in irreconcilable viewpoints encountered during scientific controversies. In this case it also reflected a deep fissure within the relativity community itself, a division that had its origin in the dual nature of the community, with one foot in physics, the other in mathematics. This dual nature, of course, had become a defining characteristic of the field during the period of its neglect by most physicists.

The Classical Theory of Fields (*Teoria Polia* in Russian) is one of the key volumes in Landau and Lifshitz's *Course of Theoretical Physics*, perhaps the most celebrated series of physics textbooks of the twentieth century. Several generations of young physicists across many countries have been shaped by them. They are noted for the depth and breadth of their coverage, a testament to Landau's brilliance in all areas of physics, and for the brevity of the writing style, the work of Lifshitz (who was one of Landau's protégés). The series was designed for the education of the best and brightest, making no concessions to the ignorance of the reader. That the one volume contained an attempt on the problem of gravitational wave emission that was technically in advance of much of the research literature of the day is perhaps an indication of the ambitious pedagogical philosophy which inspired it. In the first English edition, the case of weak gravitational fields is discussed in the following way:

> Let us consider next a weak gravitational field, produced by arbitrary bodies, moving with velocities small compared with the velocity of light.
>
> Because of the presence of matter, the equations of the gravitational field will differ from the simple wave equation of the form $\Box h_i^k = 0$ by

having, on the right side of the equality, terms coming from the energy-momentum tensor of the matter. We write these equations in the form

$$\frac{1}{2}\Box\psi_i^k = -\frac{8\pi k}{c^4}\tau_i^k \tag{6.1}$$

where we have introduced in place of the h_i^k the quantities $\psi_i^k = h_i^k - \frac{1}{2}\delta_i^k h$, where τ_i^k denotes the auxiliary quantities which are obtained upon going over from the exact equations $R_i^k - \frac{1}{2}\delta_i^k R = \frac{8\pi k}{c^4}T_i^k$ [The Einstein equations] to the case of a weak field in the approximation we are considering. It is easy to verify that the components τ_0^0 and τ_α^0 are obtained directly from the corresponding terms T_i^k by taking out from them the terms of the order of magnitude in which we are interested; as for the components τ_β^α they contain along with terms obtained from the T_β^α, also terms of second order from $R_i^k - \frac{1}{2}\delta_i^k R$ [the Einstein tensor]. (Landau and Lifshitz 1951, p. 326)

The claim here is that in place of the usual linearized wave equation, which contains no information at all about the source of the waves (plane gravitational waves in fact have no source, strictly speaking, since they do not originate at a single location), one can introduce an approximate wave equation with information concerning the material source on the right-hand side of the equation. The left-hand side retains its familiar form describing the waves themselves. The equation is limited to the case of weak fields and slow velocities, but it is asserted that the source itself can be anything that meets these criteria, including a binary star system. But the tensor construct that describes this source is pieced together by hand from the full Einstein source tensor (used in the exact Einstein equations which cannot be solved analytically for this type of system) via a complex and involved series of steps, which is laconically described in one paragraph. Higher-order terms from the exact source tensor (from the exact Einstein equations) must be dropped, and some other terms included from other parts of the Einstein field equations in order to construct the approximate source term in the correct form. The answer to the question, what is the correct form? is given at the beginning of the section on gravitational waves:

The calculation of the energy radiated by moving bodies in the form of gravitational waves requires the determination of the gravitational field in the "wave zone," i.e. at distances large compared with the wavelength of the radiated waves.

In principle, all the calculations are completely analogous to those which we carried out for electromagnetic waves. [The] equation for a weak gravitational field [the displayed equation in the previous quote] coincides in form with the equation of the retarded potentials [of electromagnetism, from an earlier section of the textbook]. Therefore we can immediately write its general solution.... (p. 329)

So if we are sufficiently dexterous, we can follow Landau and Lifshitz's recipe and construct an equation that will allow us to proceed with a radiation calculation largely analogous to the one for electromagnetic radiation damping given by them in an earlier chapter of the book. How to do this is discussed with a remarkable degree of brevity. In the second English edition (1961), the instructions on constructing the approximate wave equation with the source term have been moved to the beginning of the section on gravitational radiation damping, consolidating the argument. The entire discussion still takes little more than three pages.

What is one to make of this? Broadly speaking, one sees two reactions. One can work through the problem, decide that the procedure seems logical, and feel emboldened by the fact that the equation derived and the final result achieved agree with experience gained from the case of the electromagnetic field. One can work through it, feel shocked at the cavalier way in which terms are thrown out or included, and wind up feeling deeply suspicious that the familiar answer at the end provided the motivation behind the preceding maneuvers. If one feels this way, one will worry that here is a case where the desire to find a seemingly plausible result has directed the course of calculation with potentially fatal consequences. Everyone has had the experience of being wildly wrong precisely because of feeling so certain one was right. It happens in physics all the time. Opinions on the relevance of Landau and Lifshitz vary precisely because for some people the ease with which the seemingly analogous solution is achieved is suspicious, while for others it is, if anything, one small indication that the analysis may be correct.

The number of objections that might be made to Landau and Lifshitz's calculation is probably at least equal to the number of skeptics in this history. For instance, Landau and Lifshitz claim that the calculation is completely general. It really does not matter what motions the source makes. The motion is tied up in the carefully constructed source term. In principle it seems that the orbits of stars would produce quadrupole radiation, but what if actually solving the field equations to see how their motion is constrained produces a different result? The theory predicts that in the absence of gravitational radiation, the stars follow geodesics, which are the "straight lines," the shortest routes from here to there, of curved spacetime. The Earth follows a geodesic as it orbits the Sun. We may follow a geodesic as we walk about on the surface of the Earth taking a straight path, which is nevertheless actually curved. These geodesics certainly involve the stars in an orbit with a changing quadrupole moment. But what if allowing for the radiation alters the path of the bodies in such a way that the new, physically correct, and not approximate path produces no quadrupole radiation? Maybe the subtle interaction of orbit and radiation damping eliminates the radiation? Aha, say followers of Landau, but we know that binary star systems really do move in these geodesics to all accurate estimates. In other words, do we simply care about the physics or the internal logical argument of the theory?

One of the leading skeptics, Beck's student Peter Havas, has pointed out that in the kind of linearized approximation adopted by Landau and Lifshitz, the orbital path of a binary star is not actually a solution to the field equations at all! That is to say, that kind of gravitational bound motion is found only in solutions of the exact Einstein equations. The linearized equations do not admit of such a solution. Therefore the kind of approach taken by Landau and Lifshitz (and indeed by Einstein in his 1918 paper, in so far as anyone tried to apply it to binary stars as sources) is mathematically inconsistent. One adopts an approximation scheme but then insists on using a solution for the motion which is based on a different scheme, specifically on the exact equations in the (assumed) absence of gravitational waves. The "physicist" would think this is the right thing to do because this is more or less the way binary stars actually move, and the aim of the physicist is to simplify the mathematics while retaining as much of the physics as possible. The "mathematician" (while keeping in mind that Havas is certainly a physicist by training and inclination) would object to the obvious inconsistency in blurring two quite different methods of approximation.

In interviews with me, several relativists said they never had any doubts about the existence of gravitational waves or their emission by binary stars once they had read Landau and Lifshitz. Yet others indicated to me that

they lent no credence at all to Landau and Lifshitz. Still, those who admired
Landau and Lifshitz did not do so naively, simply because of its plausible
and familiar result. They did not regard the analogy with electromagnetism
as something to be blindly followed, and indeed, the calculation is different
from the electromagnetic case, though very similar to it at many important
points. Landau and Lifshitz themselves note one dissimilarity:

> The expression for dE/dt [given in the quote immediately below]
> contains $1/c^5$. This means that energy loss for an isolated system
> appears only in the fifth approximation in $1/c$. In the first four
> approximations the energy of the system remains constant. From
> this it follows that a system of bodies in a gravitational field can
> be described by a Lagrangian correctly to terms of the fourth order,
> v^4/c^4, in contrast to the electromagnetic field, where the Lagrangian
> exists correct only to the second order (this last is related to the fact
> that the loss of energy by electromagnetic radiation contains $1/c^3$).
> The effects caused by these auxiliary terms in the Lagrangian are,
> however, completely negligible. [The terms for the Lagrangian at
> the second-post-Newtonian approximation are presented, "without
> derivation," in a footnote.] (p. 331)

While Landau and Lifshitz may find this technical point interesting, they
dismiss it as of little overall relevance, since the extra level of the expansion
that can be treated in the gravitational case as conservative does not actually
produce measurable effects on known systems. So it is of little interest to
physics, however interesting it might be to the theorist. Indeed, since the
radiation effects occur at an even higher order, it is not surprising to read
the following statement on the same page:

> This total radiation in all directions, i.e. the energy loss of the system
> per unit time [is given by the following expression, which is the
> quadrupole formula, since $D_{\alpha\beta}$ is the quadrupole moment of the
> source, which is differentiated with respect to time thrice]

$$-\frac{dE}{dt} = \frac{k}{45c^5} \left(\frac{d^3 D_{\alpha\beta}}{dt^3} \right)^2 . \tag{6.2}$$

It is necessary to note that the numerical value of this energy loss, even
for astronomical objects, is so small that its effects on the motion,

even over cosmic time intervals, is completely negligible (thus, for double stars, energy loss in a year turns out to be 10^{-12} of total energy).

Of course, it was this fact, that there appeared to be no means of detecting the effects of gravitational radiation on any known system, which prevented any immediate controversy from erupting over Landau and Lifshitz's work, and which no doubt discouraged the authors from publishing a separate paper giving more detail about their calculation. Nevertheless, had anyone in the 1950s made a survey of the differing attitudes towards the calculation in this celebrated textbook they would have revealed the fault lines that would give birth to the long-running controversy ahead.

During World War II, work in many areas of physics was suspended, but many subjects areas were invigorated in the years immediately following, partly because physicists applied the skills and techniques learned in war work to other fields, such as in the birth of radio astronomy, and also because there was now far more money available for pure science research than had been the case before the war. Physics especially experienced a remarkable surge in the number of researchers, particularly in the United States. America became the epicenter of research in theoretical physics, not only because of the simple translation of so many key European physicists to its shores, but also because of the emergence of a brilliant generation of physicists trained there before the war. By the mid-fifties, following the tremendous success in reconciling special relativity theory with quantum theory in the form of quantum electrodynamics, the time seemed ripe to do the same for the general theory. In addition, advances in astrophysics were opening remarkable new vistas on the universe that would begin to suggest the existence of gravitational fields of previously unimagined intensity. There began to be more physicists interested in working in general relativity, and many of them could afford large groups of students and assistants to help with ambitious research programs. Although still a tiny field by the standards of other areas of physics, relativity underwent a qualitative change, from individuals or small groups of researchers to larger groups and "schools" dominated by one or two senior theorists but each involving several important young scientists. In addition, many of the new men in the field were physicists coming fresh from the battlefields of the new quantum theory with a radically different outlook to the classically and mathematically trained people who had hitherto dominated the subject.

The interesting fact about the economy of general relativity in the postwar period is the source of the field's funding. From the mid- to late fifties until the early 1970s (after the passage, in 1969, of the Mansfield Amendment, which discouraged Department of Defense funding for basic scientific research lacking direct military applications; see Kevles 1977, p. 414), much of its money came from the United States Air Force (USAF). The military of all countries had been profoundly impressed by the importance of the physicists' wartime contributions, from radar and its myriad applications to the atomic bomb itself. Since there was no way for the military to be sure where the next superweapon would come from, it made sense for them to co-opt all fields of physics, and so even highly abstract subjects like relativistic gravity began to receive military funding. The lack of comprehension of relativity that the air force labored under was well matched by the relativists' inability to fathom what it was that the air force wanted from them. This mutual misunderstanding is illustrated by the varying attitudes towards the money itself. For the air force it was surely a pittance, though I have not seen figures of the amounts involved. On the other hand, several physicists who recall the period emphasized to me that, at the time, the National Science Foundation (NSF) was seen as very much a second-best option to air force funding, which was on a more lavish scale. The NSF was itself a new postwar recreation, designed in part to offset the influence the military had gained over basic research during the war (Kevles 1977, p. 344). At least in the United States, the military took the sensible step of putting administration of the money in the hands of scientists in the field. They recruited a relativity group at the Aeronautical Research Laboratory at Wright-Patterson Air Force Base near Dayton, Ohio, and funding of outside researchers was handled by scientists at the base. Joshua Goldberg, one of the scientists involved, has written a short account of this extraordinary episode (Goldberg 1993). As he points out, the air force benefited from its association with the experts precisely to the extent that it protected itself from fraudulent claims of major scientific breakthroughs from charlatans. Many relativists speculated that the air force was deluding itself in entertaining naive expectations of antigravity paint or similar science fictional breakthroughs. Goldberg himself recalled that "one made many jokes about the interest in antigravity devices, which no doubt would be of great value to the Air Force" (Goldberg 1993, p. 100). He noted that as one of his duties was to evaluate research proposals, "there is no question I saved the air force my salary many times over" (interview).

There is no doubt that in general the military benefited from its patronage of this critical but potentially independent-minded arm of what Eisenhower called the "military-industrial-scientific" complex (I refer here to theoretical physics as a whole, rather than just to relativity). The navy originated this direct support of basic science immediately after World War II (see Kevles 1977), and according to Goldberg (1993, p. 100), the funds included some support for relativity before the advent of air force support. It was precisely to avoid the co-option of the scientific research community by the military that Senator Mike Mansfield of Montana introduced his amendment in 1969, which forbade military funding for pure science research. Henceforth, funding for relativity in the United States came primarily from the National Science Foundation (NSF), whose support up to that time had been secondary to that of the air force. The amendment also meant that groups in NATO countries could no longer benefit from American funding, as had previously been the case.

Funding clearly plays a critical role in the development of any scientific field. The field of gravitational wave theory was fortunate in this regard in that, from 1956 to 1963, Goldberg was responsible for United States Air Force support of research in general relativity. Goldberg himself was active in the study of gravitational radiation, as we shall see, and did much to encourage groups such as that of Bondi and Pirani at King's College, London. Although USAF money was available for groups outside the United States, it could not be used to support scientists based in communist countries, thus inhibiting the use of these funds to facilitate travel between the London group and Infeld's group in Warsaw, who interacted extensively (Felix Pirani, interview). The Aeronautical Research Laboratory itself was home to an active group until the 1970s. With one of his earliest grants, Goldberg used air force money to support the Chapel Hill conference organized by Bryce DeWitt, which played a role comparable to the Berne conference in the reawakening of interest in relativity theory, and about which I will have much to say in the next chapter. These two important meetings became the forerunners of the successful General Relativity and Gravitation (GR) series of conferences, which continues today. The breadth of funding possibilities available to scientists in this period is attested to by the list of other funding agencies in the Chapel Hill proceedings, which includes the National Science Foundation and the U.S. Army's Office of Ordnance Research (DeWitt 1957, p. iv).

In addition, the conference was held to inaugurate the new Institute of Field Physics at the University of North Carolina. This institute was founded with the financial support of Agnew Bahnson, a North Carolina industrialist, who had contacted both Bryce DeWitt and John Wheeler concerning the possibility of antigravity machines (Cécile DeWitt-Morette, pers. comm.). Wheeler had encouraged DeWitt to channel Bahnson's interest in more useful directions, and the DeWitts collaborated with him in establishing the institute at Chapel Hill. The conference itself, with its mixture of funding, thus represents an ideal example of the unusual military and industrial interest in gravity that emerged in the postwar period, much to the benefit of the field. Another industrialist with an interest in antigravity was Roger Babson of Massachusetts, who founded the Gravity Research Foundation in New Boston, New Hampshire. This foundation is best known for its annual prize for an essay on gravitation (for more on Babson, see Kaiser 2000). It certainly seems that the industrialists had visions of antigravity devices in mind when they began their patronage of the field. That the relativists were completely unable to fulfill the more extravagant expectations of some of their various patrons does not seem to have soured the relationship particularly, but with time, military and industrial interest in funding theoretical work in gravity faded as it became clear that no important applications were forthcoming from their work.

The Mansfield Amendment remained in force only for one year, but it helped solidify an emerging political consensus, as symbolized by the support the passage of the amendment received from both liberal and conservative congressional leaders, that basic scientific research, especially when conducted within the universities, should not be in the military's sphere, but was more appropriately the domain of civilian agencies such as the NSF (Kevles 1977, p. 414). From 1973 to very recently, the chief controller of funding for gravitation physics at the NSF had been Richard Isaacson, like Goldberg, a relativist who made important contributions to the theory of gravitational waves. Isaacson had also worked previously at the laboratory on the Wright-Patterson base. By good fortune, then, and despite the overall decrease in funding for theoretical physics precipitated by the Mansfield Amendment and the new trend in U.S. government funding, the principal source of funds for research on gravitational wave theory remained in sympathetic and knowledgeable hands.

The advantage of having an insider at the primary funding agency did not ensure that everyone in the field was sponsored to the extent that they desired or felt necessary. Complaints about the funding choices made and their effect on research directions were very noticeable on the experimental side, where groups and research programs depended very heavily on the munificence of different (usually governmental) funding agencies. But even on the theoretical side, work on the problem of motion or radiation reaction was computationally so intensive that funding for postdocs and assistants could make a big difference to a group or research program. It may be that less popular research programs (such as fast-motion approximations versus slow-motion ones, about which more later) suffered in this regard, but it is difficult to assess the extent of this factor. For every complaint made that funding was not sufficiently available to meet the needs of a given research program, there were responses that the program in question was actually not given an especially high priority by the people involved with it themselves. One must keep in mind that all of the people discussed in this book were involved in research on topics besides gravitational waves, and in most cases they are better known for their other research. Yet gravitational waves remained for many of them an interest that was pursued even in the absence of lavish funding for the subject. Bondi, one of the most famous names in the field, has said that his gravitational wave work is the research of which he is most proud (Bondi 1990, p. 79).

Whereas *from the individual perspective* the effect of funding-agency policy on research directions can appear to be malign, it nevertheless seems very likely that the subject of gravitational wave theory has benefited greatly from its close connections with the main funding agencies, especially when one keeps in mind how small and isolated a field this was at one time. Certainly Isaacson must receive a great share of the credit for the present high profile enjoyed by the field of gravitational waves, considering that LIGO is the most expensive project ever funded by the National Science Foundation.

During the 1960s, for perhaps the first time, general relativity started to become relevant to astronomy and astrophysics, leaving aside the specialist subject of cosmology. Previously, astronomy had influenced general relativity (with the Eddington expedition and the Mercury perihelion problem) rather than the other way around. This relation now began to change as a result of major transformations within each subject. The strong boost given to the practice and theory of radio observations by the military requirements of the Second World War resulted in the unexpected birth of the field of radio astronomy in the immediate postwar period (Edge and Mulkay 1976). Significantly, the new astronomical discoveries that led astronomers and astrophysicists to look towards general relativity theory for possible explanations were largely discoveries of radio astronomy. Both quasars and pulsars were types of sources for which astronomers were unprepared by their optical experience, and since both appeared to be objects that were very compact for their mass, and therefore should possess very strong gravitational fields, it was natural to turn to general relativity for a theoretical understanding of them, since it was precisely in the strong-field regime that this theory most strongly departed from the classical theory of Newton.

At the same time, perhaps partly in response to this unaccustomed interest from outside their field, relativists began to make the various predictions of the theory far more concrete than had previously been the case. The development of the idea of the black hole (and the coining of the name by Wheeler), out of the longstanding formal solution to Einstein's equations due to Schwarzschild, belongs to this period.[2] The same is true of gravitational waves, which became much more real in a theoretical context, even while they remained stubbornly outside the realm of actual detection. The texture and feel of a gravitational wave was only slowly revealed through the extraordinary work of theorists in the fifties, sixties, and seventies. Perhaps the most important single element of the convergence of relativity and astrophysics was the attempt to find an explanation for the quasi-stellar (quasar) sources. This indeed resulted in the birth of a new field, relativistic astrophysics, and a successful new series of symposia to promote it. The first of the Texas Symposia, originally promoted by Infeld's former student, Alfred Schild, was called specifically to address the quasar puzzle, and indeed, at this time gravitational waves were being presented as a component of one proposed mechanism for the power source of quasars.

This development certainly suggests a definite change in attitude towards gravitational waves, which had previously been regarded as something of a seventy-pound weakling in general relativity and as a phenomenon that would never amount to much from an experimental or observational point of view. "The weakness of the gravitational interaction makes it exceedingly unlikely that gravitational radiation will ever be the subject of direct observation," as one important researcher in the field put it (Pirani 1962, p. 199). Quasars were viewed as a problem because if their characteristically high redshifts were assessed as cosmological in origin, they must have immense, unprecedented outputs of energy from rather compact dimensions (as inferred from the variability in their luminosities on the timescale of years). That William Fowler (1964) suggested at this time that they might be powered by gravitational wave emission suggests a newfound respect for the potential of this previously unappreciated phenomenon (see also Cooperstock 1967) and also shows that astrophysicists were conscious of the potential usefulness of modern gravitational theory. It is important to keep in mind that although gravity continued to play a central role in astronomy and astrophysics, most astronomers had no training or experience with general relativity. Newtonian gravity completely sufficed for their purposes up until the 1960s, and even then it was some decades before many of them became entirely comfortable with phenomena like black holes and gravitational waves, objects constructed entirely out of pure field and containing no matter at all.

The particular topic which inaugurated the emergence of general relativity as an important ancillary field to astrophysics was that of gravitational collapse. From the early 1960s, the collapse of massive or supermassive stars, first hinted at by Chandrasekhar and investigated by Oppenheimer and his collaborators in the 1930s but largely ignored subsequently, became a favorite candidate for the power source that lay behind the apparently immense radio and optical emission of quasars. In 1963 Peter Bergmann (one of Einstein's assistants from the thirties and the founder of the first major school of general relativity) and Alfred Schild (one of Infeld's students at Toronto) issued a call for a symposium to be held on this subject, one of its goals being efforts to avoid the catastrophe of gravitational collapse to a singularity, in which the entire mass of a star would be concentrated at a single infinitesimally small point of space (recall Einstein and Rosen's claim that the existence of a singularity invalidated the relativistic solution for gravitational plane waves). The symposium was actually held in late 1963 at the University of Texas, where Schild worked. It subsequently gave birth to a regular and highly successful series known collectively as the Texas Symposia on Relativistic Astrophysics, the name of the field said to have been coined by Schild (Ehlers 1980, introduction). From the first meeting, beginning with Fowler's article (1964) on the role for gravitational radiation in gravitational collapse and quasar emission, the topic of gravitational waves was invariably addressed at each symposium. Therefore, this phenomenon was clearly viewed as one of the important elements of the new subject.

The accidental nature of the rapid development of the field of relativistic astrophysics is illustrated by an excerpt from Gold's speech at the closing of the first Texas symposium:

> Here we have a case that allowed one to suggest that the relativists with their sophisticated work were not only magnificent cultural ornaments but might actually be useful in science! Everyone is pleased: the relativists who feel they are being appreciated, who are suddenly experts in a field they hardly knew existed; the astrophysicists for having enlarged their domain, by the annexation of another subject— general relativity. It is all very pleasing, so let us hope that it is right. What a shame it would be if we had to go and dismiss all the relativists again. (Robinson, Schild, and Schucking 1965, p. 470)

But, in spite of Gold's anxiety, the genie was out of the bottle. General relativity was extremely slow to shed much light on the topic of quasars, but the reorientation that was encouraged within relativity and the success of the great body of work that went forward on gravitational collapse and black holes invigorated the subject as never before and lent great vitality to the idea of mixing relativity with astrophysics (for the history of general relativity and black holes, see an authoritative popular account in Thorne 1994 and the definitive historical discussion in Eisenstaedt 2006). Pulsars, discovered in 1967 by Jocelyn Bell and Tony Hewish, were rather quickly identified with the idea of neutron stars, which had been knocking around in the background since the thirties and greatly helped to underpin relativity's new role. Gravitational waves also played a role in the subsequent development of pulsar theory, since their emission was expected to quickly damp the wild pulsations of the neutron core at the end of the gravitational collapse of the parent star (Thorne 1969). The extent to which the internal dynamic within general relativity, encouraged by the move towards astrophysics, developed its own momentum is evidenced by the growth of interest in gravitational waves, which preceded by some time the emergence of any observational input into the subject. By 1967 there was a new sentiment about the prospects for gravitational wave detection, characterized by Wheeler's statement that "gravitational waves, I cannot help but feel, are going to be one of the big discoveries of the next ten years. One will detect them for the first time. That is one great prediction of Einstein's theory" (Wheeler 1967).

Still, it was to the great surprise of most theorists that Joseph Weber announced in 1969 that he was detecting gravitational waves (Weber 1969). Although his results, which confounded all theoretical predictions of source strengths then and since, were eventually discounted amid much controversy, they focused much attention on the subject and sparked a great increase in the number of experimentalists working on gravitational waves (see Collins 2004 for a detailed account and Franklin 1994 for an alternative viewpoint). On the theoretical front, research in the 1960s on black holes, cosmology, and other topics had made the field of relativity very relevant to astrophysics. Gravitational waves shared somewhat in this popularity and seemed likely to continue to grow in practical importance as experimental interest waxed. The discovery of the first binary pulsar, PSR 1913 + 16 (Hulse and Taylor 1975), was the fortunate and serendipitous occasion that sealed this promise. Over a period of several years Joseph Taylor and his collaborators were able to test a number of predictions of general relativity through their measurements of the orbit of the binary pulsar, culminating, in late 1978, in their claim that there was an apparent decay in the orbital period of the binary system, which was in accord with the predictions of Einstein's quadrupole formula. This announcement, fittingly, came at the Nineth Texas Symposium on Astrophysics in Munich, Germany. The important role of conferences in the development of this field is illustrated again in the next chapter, where we look at the first important relativity conference to be held in America.

7

Debating the Analogy

So in the years immediately after World War II, until after the Berne jubilee conference of 1955, general relativity was at a low ebb (Eisenstaedt 1986a, 1986b). What work there was on the radiation problem seemed confused and controversial. It must have seemed that a good part of the blame for this state of affairs lay with the relative scarcity of researchers in this field. There is, after all, a nonlinear affect at work here, since most of the time it is not just numbers of researchers that are important in science, but also the number of links between researchers. It is well known that as the number of people in a network increases linearly, the number of possible links in the network increases geometrically. Whether the network will really be maximally linked is, of course, a key question, but it must be admitted that the situation which existed in general relativity until the midfifties (at least) was certain to minimize the number of links between researchers. When there is only a small number of researchers, when they are geographically located in many different countries (research in general relativity has always had a very international profile, a potential source of both strength and weakness), and when they are to be found working in or hailing from different disciplines (some in mathematics; some in physics; some even trained in chemistry, such as Richard Chase Tolman; some interested in mathematical physics; some in cosmology; and all lacking dedicated journals to publish their results where they would be likely to attract the attention of like-minded researchers), one necessarily has a weak network. We have already witnessed the characteristics of a weak network: the needless repetition of previous work, which has been lost, forgotten or ignored; the lack of any follow-up to important questions or difficulties raised by existing work; the absence of any consensus as to the definition of important concepts, such as spacetime singularities. Given all this, it might have been expected that the level of confusion surrounding the emission of gravitational waves from binary stars would steadily decrease with time as the renaissance of general relativity got under way. Instead, the level of confusion generally increased, at least if one goes by the number of possible answers to this question that were reported.

A great deal of the confusion arose out of attempts to calculate, via the problem of motion, the rate of energy loss in binary star systems due to gravitational wave emission, as Eddington had done for the case of a rotating rod. In this approach one iteratively solves the equations of motion for the system, rather than simply calculating the energy carried away by gravitational waves, as Einstein and Eddington had done when they first derived the quadrupole formula. Between 1947 and 1970, dozens of different calculations by a wide variety of methods gave an almost equally wide variety of results. Not only did many calculations not recover the quadrupole formula at the leading order, but some (principally those due to Infeld and collaborators) found no emission, and three separate publications (Hu 1947; Peres 1959; Smith and Havas 1965) found that the binary would paradoxically gain energy as the result of emitting gravitational waves. Confounding the issue further was the problem of coming up with an unambiguous way of defining the energy carried by the waves themselves. Right away, as interest in general relativity began to recover after 1955, it was realized that such a definition was not a trivial issue.

If the initial consequence of more researchers taking up the problems of gravitational waves was more confusion, or at least confusion better articulated than the silent ignorance that had preceded it, more vigorous pursuit of the questions at hand was nevertheless a necessary first step to their solution. The initial revival of interest in general relativity, at least in the minds of many of the relativists themselves, took place at the Berne conference of 1955. It seems to have been a surprise to some that other physicists and mathematicians actually shared their interest in this difficult and esoteric theory. Hermann Bondi, who had attended this conference and was to be one of the central figures in gravitational wave theory for the better part of the following decade, recalled that at this conference there were several discussions concerning whether gravitational waves existed. He himself, on the lookout for a research topic for his young research group at King's College, London, that was "a subject not widely pursued at the time because we could not compete with the big battalions," responded to a suggestion made to him by Marcus Fierz, another leading Swiss physicist. Bondi recalled:

> Discussions were good, and puzzlement about what the theory [of general relativity] said about gravitational waves was still prevalent at the time. . . . At the end of one such discussion, [Fierz] turned to me and said, "The problem of gravitational waves in general relativity is now ripe for solution and you are the person to solve it."(Bondi 1987, p. 17)

At the time of the Berne conference, the two relativists who were most vocal in their skepticism of gravitational waves were Rosen and Infeld, who provided a direct link to the prewar period when Einstein had made his last contribution to the subject. As 1955 was the year of Einstein's death, he did not live to witness the revival of interest in his theory, though with his characteristic independence he had made light of, and even savored, his own intellectual isolation in his later years. Following his Soviet adventure, Rosen had returned to America before the war, and after the end of the war emigrated to Israel, to the Technion Institute at Haifa. Rosen's contribution to renewing the debate on gravitational waves was made at the Berne conference itself and is worth discussing in some detail. He returned to the cylindrical wave solution of his 1937 paper with Einstein, suggesting that gravitational waves cannot actually transport energy (Rosen 1955).

It is a feature of general relativity that the energy contained in the gravitational field, and thus the energy in gravitational radiation, is not described in a coordinate-invariant way. This energy can be thought of as real enough and can be converted into other forms of energy which can be expressed invariantly, but the principle of equivalence demands that gravitational field energy itself not be represented in an invariant way. The reason is that any observer in a gravitational field is always entitled to imagine herself as if she were falling freely, experiencing no weight. This, indeed, was the insight upon which Einstein had founded the theory, the principle of equivalence, which says no one can tell the difference between being accelerated and being in a gravitational field. Thus one is always free to imagine that gravitational field energy is really in another form, such as kinetic energy, the energy associated with motion. One can make gravitational fields disappear by adopting the appropriate *point of view*. One does not necessarily transform away the entire field energy of a planet, but one can always choose coordinates on a small portion of its surface so as to eliminate the field energy in that region (recall here Nordström's puzzlement at his ability to transform gravitational field energy away, which he expressed in his correspondence with Einstein; see chapter 3). Thus it is said that gravitational field energy is *nonlocalizable*: you can't point to a physical location at which is resides; it all depends on how you look at it.

This problem of the nonlocalizability of field energy had been recognized in the very early days of relativity. The phrasing was already used in early papers by Eddington and Wolfgang Pauli, for instance, and the whole idea began with Einstein, who was obliged to fiercely defend his views against prominent theorists and mathematicians (most prominent among them the Italian mathematician, Tullio Levi-Civita), who thought his definition of field energy wrongheaded. He employed the infamous pseudo-tensor to describe it, the name indicating precisely that its values changed as one changed coordinates (tensorial quantities are said to be invariant to coordinate transformations). The Einstein energy pseudo-tensor, and various other definitions of pseudo-tensors, continued to play a central role in gravitational field theory as a way of describing the flux of energy carried by the waves, in spite of the spate of early objections from Levi-Civita and others (see Cattani and De Maria 1993 for a discussion).

So in 1955, Rosen observed that two different definitions of the pseudo-tensor, Einstein's original and that used by Landau and Lifschitz, appeared to show that gravitational waves carried no energy at all, in the sense that terms related to energy transport were zero in both cases when applied to the exact cylindrical waves from his paper with Einstein. Although drawing conclusions on the tentative basis of the pseudo-tensor was regarded as dangerous, Rosen observed that the result seemed to support the views of Infeld and his collaborators who, as we shall see, were arguing that gravitational waves did not carry energy away from systems like binary stars. It was not long before it was shown in 1959 by John Stachel (who many years later was to be the first editor of the *Collected Papers of Albert Einstein*) that Rosen's no-energy result was coordinate dependent. What was most apparent was that the notion of energy transport in gravitational waves was not well understood.

Two already distinguished theorists who moved into general relativity in the late fifties were Hermann Bondi and John Wheeler. Though young, Wheeler was already a leading figure in theoretical physics in the United States, where the discipline was still very young, having really only emerged as a specialty after the First World War (prior to that pure theorists were a rarity in American physics). In the midfifties Wheeler, who already had a well-established reputation in nuclear physics and classical electrodynamics, became interested in general relativity. The enormous success of theoretical physics in the postwar period, especially in the development of quantum field theory spurred interest in the most elegant of all field theories, especially in the question of how to quantize this quintessentially classical theory. Hermann Bondi, an Austrian physicist who had settled in England and who was one of the founders of the steady-state theory of cosmology (at that time a serious rival to the emerging big bang model of the early universe), was inspired to work on gravitational waves by what he heard at the Berne conference. Having recently taken up a position at King's College, London, Bondi was in the process of forming, with C. W. Kilmister and (a little later) Felix Pirani, one of the most influential relativity groups during the renaissance of the subject. The "big battalions," whose competition Bondi feared, would come to be represented within the field of relativity by Wheeler's group. From the early 1950s, when Wheeler's interest in the theory blossomed, a steady stream of students ensured that Wheeler's would be one of the two leading schools of relativity theory in the second half of the twentieth century, along with one of the earliest of the relativity schools—that of Einstein's former assistant, Peter Bergmann.

While Rosen's doubts about whether gravitational waves could transmit energy were raised at the Berne conference, they were answered, as far as the question of whether absorbers of gravitational wave energy could exist, at the next important conference on gravity, which took place in 1957. Whereas the Berne conference was held partly to commemorate relativity's past, the first major conference held specifically to discuss recent work in general relativity took place in 1957 at Chapel Hill, North Carolina. The conference organizers were a husband and wife team, Cécile DeWitt-Morette and Bryce DeWitt, two important relativists in this period who, among other things, greatly advanced the understanding of how electromagnetic waves propagate through the curved spacetime of general relativity. As we have seen, in organizing the conference, they benefited from the new sources of funding available at this time.

The Berne conference had been attended by only nine America-based physicists, out of a total participation of ninety or so (Mercier and Kervaire 1956). The Chapel Hill conference, therefore, did for the American relativity community what Berne had done for the Europeans, besides being more open to students and younger researchers. At Chapel Hill, fully thirteen out of forty-five participants were based outside America, and support from the air force was to help greatly in facilitating further contact between the relativists on both continents. The USAF made the Military Air Transport Service available for foreigners attending the Chapel Hill conference and also for Americans attending subsequent gravity meetings in Europe (up until 1965; Goldberg 1993, pp. 90–91). What was required was an organization to make sure that the conferences themselves would continue to take place. After the Berne and Chapel Hill conferences had given both European and American relativists a taste for like-minded companionship, an organization, the International Committee on General Relativity and Gravitation, was developed to encourage a permanent series of conferences, known as the GR series of conferences (conferences on general relativity and gravitation). The first GR conference was held in 1959 at Royaumont in France, but both Chapel Hill and Berne have been referred to as GR0, for their inspirational role in the founding of these international conferences. Although the Chapel Hill conference was attended by only forty-five physicists, according to the proceedings, recent meetings in this series are enormous affairs, with many hundreds of attendees (600 abstracts for talks were submitted to the GR17 conference in Dublin in 2004). The International Committee also founded a journal, *General Relativity and Gravitation*, to provide a venue for papers which would guarantee an interested audience.

As at Berne, the existence of gravitational waves was a lively topic of debate at Chapel Hill. Goldberg, who had obviously played a key role in bringing the conference about, recalled:

> It was the first . . . conference in which postwar students of general relativity were able to participate, and it was a marvelous experience for us. . . . My most vivid recollection is long discussions with Bondi and Gold [Thomas Gold, like Bondi an Austrian émigré, was

Figure 7.1. Tommy Gold and Hermann Bondi at an International Astronomical Union meeting in 1952, with their collaborator on the steady-state theory, Fred Hoyle. Gold and Bondi began as skeptics on the existence of gravitational waves, but Bondi ended up playing a key role in showing that they exist. (By permission of the Master and Fellows of St. John's College, Cambridge University)

one of the cofounders, with Bondi and the English astronomer Fred Hoyle of steady-state cosmology], who took the position that gravitational radiation did not exist. I no longer recall the arguments they used, but I presume that those discussions prompted Bondi to develop his analysis at null infinity. Obviously, he had to show mathematically that reasonable boundary conditions inhibited gravitational radiation. (Goldberg 1993, p. 91)

Despite Bondi's skepticism regarding the reality of gravitational waves, it was he who was to give the most authoritative response to Rosen's challenge from the Berne conference. The key contribution that enabled Bondi's reply to Rosen, and which was itself announced at Chapel Hill, was due to Felix Pirani, Bondi's new colleague at King's College, London. Pirani, who had begun his education in relativity in Canada with Infeld's group at the University of Toronto before moving first to the United States and then to England, had been influenced by the Irish mathematical relativist

John Synge during a year at the Dublin Institute of Advanced Studies. In the very early days of relativity Synge had already drawn attention to the importance of the equation of geodesic deviation in general relativity. This equation is essentially a local way of measuring curvature of spacetime. One can measure the local geometry of the spacetime location one finds oneself in by observing its geometry. In flat space, as in flat spacetime, parallel straight lines, known as geodesics, do not converge or diverge from each other. As stated in one of Euclid's most famous axioms, they never cross, even when projected all the way to infinity. This is not true in curved space or spacetime, as was realized by mathematicians in the nineteenth century. In curved space, geodesics (the shortest route between two points) are not straight, and parallel geodesics may cross each other and often do (just in the way that lines of longitude on the earth, though apparently parallel at any given location, actually all cross each other at the poles). Since moving particles in general relativity travel on geodesics of the local spacetime (in the absence, at any rate, of forces other than gravity, and in the absence of gravitational radiation), Synge recognized that the motion of particles could be used to measure the local curvature of spacetime. Pirani realized that this would be a very natural way to test for the presence of gravitational waves.

If we return for a minute to the seaweed analogy, we can think about the case where the oceanography professor is in mid-ocean, trying to observe the effect of waves on seaweed from aboard ship. Since the shore is not in view, he has difficulty in using a land-centered reference frame. Of course he could try and create an invisible set of coordinates not tied to the motion of the sea, and this is indeed what navigators from the age of exploration on had tried to do in measuring longitude. But this would be a most unsatisfactory way of observing wave motion. All he need do is place two corks close together in the water and see if they move closer to and farther away from each other in a rhythmic fashion. Such motion would be a clear indication of the presence of ocean waves, and a very similar effect could be observable in particles floating in free space if a gravitational wave passed by.

Pirani proposed that the *equation of geodesic deviation* would be the most appropriate measure of the effect of a gravitational wave, rather than an attempt to define and locate the energy in the wave:

By measurements of the relative accelerations of several different pairs of particles, one may obtain full details about the Riemann tensor.

One can thus very easily imagine an experiment for measuring the physical components of the Riemann tensor. (DeWitt 1957, p. 61)[1]

In response, Bondi asked:

Can one construct in this way an absorber for gravitational energy by inserting a . . . term, to learn what part of the Riemann tensor would be the energy-producing one, because it is that part that we want to isolate to study gravitational waves?

And to that Pirani responded that one could do so easily. Bondi shortly afterwards published a letter in *Nature* that advanced a famous thought experiment based upon Pirani's proposal. In Bondi's thought experiment he imagined a stick with some small rings fitted around it so that they could slide up and down, but not without friction against the stick. If a gravitational wave passed by, the changing curvature produced by it would shrink and decrease the distance between the rings, but the stick, which was a solid object held together by electromagnetic forces, would be expected to resist this alteration in distance. Thus, from the observer's point of view, we would see the rings move up and down along the stick, and in doing so friction between them and the stick would generate heat (see figure 7.2). Where would the energy for this heat come from? Since the receiving system is inert, it could only have come from the gravitational wave, which therefore must carry energy, regardless of what the pseudo-tensor says.

By this time, in any case, it was becoming apparent that the energy at any given point in the wave was an entirely coordinate-dependent thing. It might be zero, or nonzero, depending on the coordinates used (Stachel 1959), which merely showed that this kind of energy was nonlocal, in the sense that one could move it around by just looking at from a different viewpoint. For the moment, because there appeared no easy way to analyze the energy in the wave, skepticism won out. It would clearly be better to deal with an invariant quantity, such as the spacetime curvature, or else to analyze the source or the receiver to see if energy had been lost or gained due to radiation. If the source or the receiver could be treated as a system in isolation from the rest of the universe (a star very far from other stars, for instance) then it seemed plausible there would be ways to define its mass-energy in a coordinate-independent way.

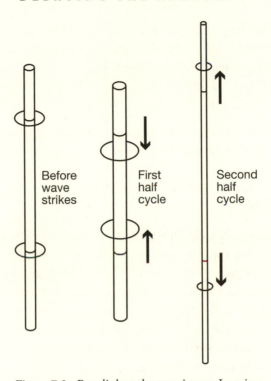

Before
wave
strikes

First
half
cycle

Second
half
cycle

Figure 7.2. Bondi thought experiment. Imagine a stick with rings that are looped around it but free to slide up and down against it. If a gravitational wave passes by, it will alternately stretch and shrink the stick, and will cause the rings to move also. But if the stick is held together by internal forces that resist this motion, then we expect the rings to respond more completely to the wave and thus slide along the stick from their original (marked) positions. If there is friction between the sliding rings and the stick, heat energy will be produced. Note the characteristic quadrupolar beam pattern of a gravitational wave. An object in a gravitational wave is shortened and fattened in the first part of a cycle, and stretched and thinned in the second part.

Later in the Chapel Hill conference another elaboration of the same thought experiment idea took place during one of the sessions on the quantization of gravity. Following Richard Feynman's presentation on the need for a quantum theory of gravity, one of the pioneers of quantum gravity, Leon Rosenfeld, made the following remark:

"It seems to me that the question of the existence and absorption of waves is crucial for the question whether there is any meaning in quantizing gravitation. In electrodynamics the whole idea of quantization comes from the radiation field."(DeWitt 1957, p. 141; quotation marks in original)

We see here how the old analogy with electromagnetism played a central role in reviving interest in gravitational waves at this time. The chief expectation of the physicists who turned to work on general relativity in the fifties was that it would be a stepping stone to a new quantum field theory of gravity. Since it was the study of electromagnetic radiation by Planck and Einstein that had originally led to the discovery of the old quantum theory and had later played a central role also in the breakthrough to the new quantum mechanics, it seemed reasonable to study gravitational waves with a view to finding a path to quantum gravity. Certainly the largest single topic at the Chapel Hill conference was that of quantum gravity. While classical general relativity was becoming more relevant all the time to astrophysics, it was the expectation of many physicists that the theory was on the point of disappearing altogether, to be subsumed in a new quantum field theory. Whereas it had been Einstein's hope that a unified field theory would be classical, explaining away quantum mechanics by some mechanism which avoided probablistic interpretations of the motion of individual particles, the vast majority of physicists in the postwar period expected that classical general relativity must conform itself to the curious laws of the quantum world.

Feynman, whose presentation had included thought experiments which sought to demonstrate that the classical theory of gravity could not coexist with a quantum theory of electromagnetism, agreed that some of his arguments in favor of quantization depended on the existence of waves. Bondi responded that "this vexed question of the existence of gravitational waves does become more important for this reason"(p. 142). After all, if gravitational waves did not exist, then perhaps this was a clue that gravity was fundamentally different from other fields and might actually not be a quantum theory at all. Feynman then remarked that Pirani's earlier presentation seemed to provide the answer. Appealing to the equation of geodesic deviation, he argued that a particle lying beside a stick would be rubbed back and forth against the stick by a passing wave, and the friction would generate heat, so that energy would have been extracted from the wave. Furthermore, he thought that any system which could be an absorber of waves could also be an emitter. For these reasons, he expected gravitational waves to exist, concluding, "I hesitated to say all this because I don't know if this was all known as I wasn't here at the session on gravity waves" (supplement to De Witt 1957, p. 143).

In fact, in his anecdotal memoir *Surely You're Joking, Mr. Feynman*, we learn that Feynman was a day late getting to the conference. Arriving at the airport he did not know precisely where the conference was being held (the center at the Chapel Hill campus that was organizing the conference was then brand new). As there were two nearby universities he was faced with a problem at the taxi rank, since none of the other participants were available to help with directions:

> I had nothing with me that showed which one it was, and there was nobody else going to the conference a day late like I was.
>
> That gave me an idea. "Listen," I said to the dispatcher. "The main meeting began yesterday, so there were a whole lot of guys going to the meeting who must have come through here yesterday. Let me describe them to you: They would have their heads kind of in the air, and they would be talking to each other, not paying attention to where they were going, saying things to each other, like G-mu-nu G-mu-nu" [This might refer either to the Einstein tensor, which is usually written $G_{\mu\nu}$, or to the metric tensor, written $g_{\mu\nu}$.]
>
> His face lit up. "Ah, yes," he said. "You mean Chapel Hill!" He called the next taxi waiting in line. "Take this man to the university at Chapel Hill."
>
> "Thank you," I said, and I went to the conference. (Feynman 1984, pp. 258–259)

Now there is one objection that has occasionally been made to Feynman's thought experiment, even down to the present day.[2] It actually is relevant to many or most proposed methods of detecting gravitational waves and is another illustration of the difficulty of proving that gravitational waves are capable of carrying energy from and to physical systems. Feynman actually addressed it during his talk:

> I heard the objection that maybe the gravity field makes the stick expand and contract too in such a way that there is no relative motion of particle and stick. But this cannot be. Since the amplitude of [one particle]'s motion is proportional to the distance from the [other particle], to compensate it the stick would have to stretch and shorten by certain ratios of its own length. Yet at the center it does no such thing, for it is in the natural metric [Feynman's term for the coordinate system in which the stick and the particle which is actually attached to it feel no gravitational field]—and that means that the lengths determined by size of atoms etc. are correct and unchanging at the origin. . . . I think that any changes in the rod lengths would [be at least two orders of magnitude less than the changes in distance between the particles] so surely the masses would rub the rod. (Supplement to DeWitt 1957, p. 143)

Keep in mind that the stick is held together by forces between the atoms, which attempt to oppose the stretching or contracting force exerted upon parts of the stick by the gravitational wave. If we think of the stick as fairly rigid, then it is logical to assume that it will respond less completely to the wave than the sliding particles will. If the stick is elastic, then it may move with the sliders to begin with, but when the wave has passed by, the sliders will stop moving, while the stretched stick should snap back and vibrate back and forth. In the elastic case it is the stick that rubs up against the particles, and then only after the wave has passed by, but either way, some energy transfer is possible.

Feynman also went on to discuss the case of the energy radiated by a "double star." We know that he did calculations on this topic from a letter he wrote to the physicist Victor Weisskopf in 1961. In this case it is unlikely that he convinced Bondi and the other skeptics. As we shall see, the question of whether binary stars radiate gravitational waves was still being discussed five years later at a subsequent conference on relativity, and that would prove irksome to Feynman. As he wrote to Weisskopf, who had asked about Feynman's work on gravitational waves, Feynman went into some detail about radiation reaction and concluded:

> In fact it was this entire argument used in reverse that I made at a conference in North Carolina several years ago to convince people that gravity waves must carry energy (for they generate heat by rubbing the particles against the rods). . . . I was surprised to find a whole day at the conference devoted to this question, and that "experts" were confused. That is what comes from looking for conserved energy tensors, etc. instead of asking "can the waves do work?" I know it will amuse you to see how "Weisskopfian" (i.e. physical) thinking is so useful here.
>
> Well, this ought to hold you for a while. If you have any more questions let me know. How can we detect the waves? Be careful of people who look for slight delay phase lags inside the wave zone. Just as in electricity only still more completely, the total field conspires to look instantaneous within distances less than a wave length to a very high order of accuracy. Only beyond the wave length can a clear proof of waves be found. I have not seen any plans for any such experiments, except by crackpots.
>
> And even further, how can we experimentally verify that these waves are quantized? Maybe they are not. Maybe gravity is a way that quantum mechanics fails at large distances. Isn't it interesting to live in our time and have such wonderful puzzles to work on? (Feynman Papers, California Institute of Technology, box 29, folder 14)[3]

Pirani's work not only influenced Bondi and Feynman but also Joseph Weber, an engineer-turned-physicist who was considering the idea of trying to experimentally detect gravitational radiation (those interested in Weber's extraordinary career should read Collins 2004). In the years after Chapel Hill, Weber designed a device for measuring the Riemann curvature of spacetime when it changed dynamically, as in a gravitational wave. Immediately after Chapel Hill, Weber, who was at that time working in John Wheeler's group in Princeton, wrote with Wheeler a paper (Weber and Wheeler 1957) that was the third example of a thought experiment based on Pirani's work and that may with some justification be seen as the starting point in the history of the experimental detection of gravitational waves. It is a testament to the immediacy of science that an experimentalist was

Figure 7.3. John Wheeler, one of the
more creative physicists of the twentieth century. His influence on the subject of gravitational waves came about
mostly through the brilliant series of
students he mentored. (Courtesy AIP
Emilio Segré Visual Archives, Physics
Today Collection)

present and contributing to the debate as to whether gravitational waves
could exist and simply went out to try to detect them. In many such cases
the theoretical debate has been preempted by an early experimental result.
In this case the detection effort proved such a monumental challenge, with
still no confirmed detection as of this writing, that the theorists had free
rein to continue the debate almost indefinitely.

To whom should we assign priority as the originator of this famous
thought experiment? It is generally, and with good reason, given to Bondi
as the first to publish the result, but it has been argued that the experiment, which is usually referred to as the Bondi thought experiment, would
be better named the Bondi-Feynman thought experiment, as Feynman
gave such a clear exposition of it at the conference, according to a supplement to the official transcript of the proceedings. Certainly it is only

right to admire Feynman's quickness of thought in presenting the thought experiment towards the end of the conference, in response to the skepticism regarding the existence of gravitational waves. But that Bondi had the same thought almost as immediately is also preserved in the transcripts, for as we have seen, his first comment after Pirani's presentation zeros in on the possibility of proving that absorbers of energy from gravitational waves can exist. It is also likely that Wheeler and Weber came up with the idea independently. As Bondi himself remarked to me, "I knew it was important, I thought it was obvious" (interview). Pirani's work in finding the right tool had certainly made it so, at least to minds of the caliber of Bondi's, Feynman's, Wheeler's, and Weber's.

As we saw in the introduction, Bondi's skepticism concerning gravitational waves and the analogy with electromagnetism that underpinned it was matched by John Wheeler's fidelity to the analogy. This divergence in outlook was a symptom of the difficulties that would be experienced by the new field (nearly a half century old, strictly speaking, yet greatly underdeveloped, as we have seen), into which researchers from diverse backgrounds were beginning to migrate. Bondi and Wheeler, indeed, were rather closer in background than many of the other newcomers to the field. In spite of their differing views on the value of the electromagnetic analogy, Bondi and Wheeler each realized that Pirani's work showed how a thought experiment could be constructed to show that gravitational waves must carry energy. Bondi and Wheeler were similar in that they chose general relativity as the field in which they wished to make a second career and each became devoted to the peculiar mores of that field. Wheeler's former student Feynman was an example of a physicist approaching relativity theory with the intention of correcting the historical accident that had led Einstein to construct it differently from the way field theory had developed in the previous half century.

Nevertheless, Bondi, Feynman, and Wheeler each arrived in an instant at the same place in regard to the "Bondi-Feynman" thought experiment, which played such a key role in persuading physicists that gravitational waves really must exist. The genius of physics is precisely its convergent nature, wherein three people coming from very different places arrive at the same result. Many fields are divergent, in that people sharing common backgrounds often end up disagreeing over critical results. It is worth noting that in this particular case it is certainly not the rigor of experimental verification that was the engine of this convergence. There were no experiments on gravitational waves, and Weber's efforts to provide them would not bear any fruit for a long time. If contact with the natural world is invoked as source of this convergent tendency, the best that we can say is that analogy enabled each of the three to make use of experience gained from other experiments to arrive at the "right" conclusion.

However convergent physics as a discipline may be, it is hardly free from disagreement. The question of whether gravitational waves carry energy is a nice example of how opinions in physics do not agree universally upon the one point. Following the thought experiment as enunciated by Bondi and Feynman in 1957, very few relativists doubted that gravitational waves transported energy in a manner essentially analogous to electromagnetic waves, although it was accepted that finding an invariant way to calculate how much energy was being propagated was another matter entirely. It was understood that even though the energy was described by the noninvariant pseudo-tensor, that fact was not a good argument for concluding that the energy did not exist, and attention shifted to the study of changing curvature very far from a source as the signature characteristic of gravitational waves. It was only with the work of Richard Isaacson, a student of Charles Misner, who was one of the early products of Wheeler's group after its switch to relativity, that it was shown how, by averaging the energy in the wave over several wavelengths of the wave, the energy in a gravitational wave could be described in a coordinate-invariant way. Although a change of coordinates can shift the energy from one part of the wave to another, the energy must generally remain somewhere within a wavelength or so of where it started. Isaacson's invariant representation of wave energy was enthusiastically taken up and became a central tool in research related to the actual detection of gravitational waves. Yet even in the 1990s there have been counterarguments by at least one relativist, the Canadian physicist Fred Cooperstock, that gravitational waves do not carry energy and that there is a flaw in the original Bondi-Feynman thought experiment. Cooperstock and others have even argued that the kind of gravitational wave detector originally built by Joe Weber cannot possibly work because of this fault. Nevertheless it appears the hour for debate on this topic has long since passed, and there has been no recent controversy surrounding it. As we shall see, once a critical mass of researchers in a field regard a problem

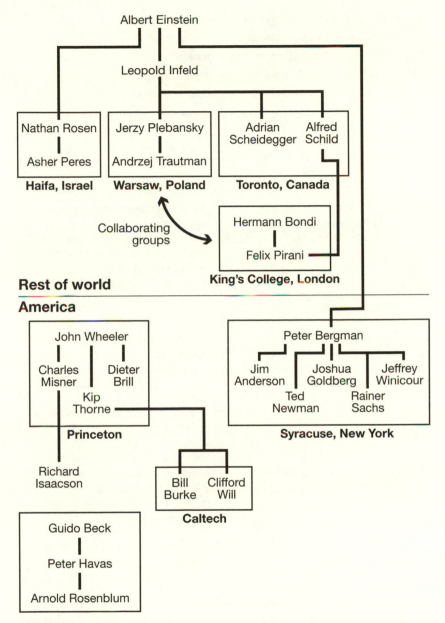

Figure 7.4. Academic family tree. Links represent a mentor and his student or (in the case of Einstein) his assistant. Names inside closed curves are members of groups or schools.

Calculation of Energy Carried in the Wave

1914 Abraham realizes dipole gravitational waves do not exist.

1916 Einstein introduces first theory of gravitational waves.

1918 Einstein first calculates the quadrupole formula.

1922 Eddington corrects a factor of 2 error in Einstein's formula.

Calculation of Problem of Motion of a Wave Source

1936 Einstein and Rosen try to prove that gravity waves do not exist.

1923 Eddington shows that the quadrupole formula applies to a spinning rod with a finite speed of propagation of gravity.

1938 Einstein Infeld and Hoffman (EIH) do first post-Newtonian order problem of motion.

1941 Landau and Lifshitz claim the quadrupole formula applies to waves from binary stars.

1947 Ning Hu applies EIH to case of emission of waves from a binary star but finds energy gain.

1953 Infeld and Scheidegger claim that binary stars do not emit gravitational waves.

1955 Rosen argues that gravitational waves cannot carry energy.

1955 Fock recovers the quadrupole formula using his own slow-motion calculation.

1958 Trautman shows that energy loss by binary stars cannot be transformed away.

1957 Bondi and others give thought experiment to show that gravity waves do carry energy.

1959 Peres finds energy gain in binaries.

1959,60 Peres discovers that his previous result was due to incorrect boundary conditions. New result agrees with quadrupole formula.

1962 Bondi et al. show that systems which emit gravitational waves lost mass in doing so.

1960 Infeld and Plebansky textbook, in which Infeld still argues that binary stars do undergo radiation damping.

1964 Brill and Hartle show that the gravitational geon has mass.

1965 Smith and Havas use fast motion calculation but find energy gain by binary system emitting waves.

Figure 7.5. History of the Quadrupole Formula, up to 1970. Entries in italics refer to a calculation obtaining a zero or negative energy result for gravitational waves emitted by a binary star.

1969 Burke introduces matched asymptotic expansions into general relativity, permitting consistent matching between near field (source) and far field (radiation) calculations.

1979 Rosen argues from absorber theory that binary stars do not lose energy to emitted waves.

1970 Chandrasekhar and Esposito recover quadrupole formula for binary star system with stars modeled as more physically realistic extended bodies.

Figure 7.5. (Continued)

as settled, no amount of effort by a small minority can reopen it, unless in unusual circumstances.

After 1957 attention shifted away from the old question, do gravitational wave exist at all? to a much more difficult one, do binary star systems radiate them and, in doing so, lose energy and decay in their orbits? Such behavior had never been observed, and it was clear that the effect was too tiny to see in all known binaries. Theoretically this question was part of the problem of motion in gravitationally bound systems, and to follow that story, we must return to Einstein and his collaborators and their work in the late thirties.

8

The Problem of Motion

Einstein and Rosen's abortive effort to disprove the existence of gravitational waves was followed by a project upon which Einstein and Infeld embarked together with another of Einstein's younger collaborators, Banesh Hoffmann. They wished to develop the post-Newtonian theory of the problem of motion, an ambitious project involving intensive calculations written up by Einstein, Infeld, and Hoffmann in a 1938 paper, a paper so well known to relativists that it is always referred to by the initials of its authors, EIH. In pursuing the EIH research, Einstein wished to vindicate an earlier conjecture (Einstein and Grommer 1927) that in general relativity the motions which particles are permitted to make are completely determined by the field equations, in contrast to other field theories where a separate force law is invoked. For instance, in the case of electromagnetism, the field equations of Maxwell describe what force a charged particle, such as an electron, feels in the presence of other charged particles, but Newton's second law must be used to say how the particle will move in response to this force. Since that motion will in turn affect the field produced by the particle, we see there is an interplay between the field equations and the force law such that each is necessary to solve the problem. In a paper of 1927 with his then assistant Jacob Grommer, Einstein had discussed the interesting property of general relativity, that its field equations completely determine the motion of particles in a gravitational field, but that was something which he and others had appreciated about the theory from the beginning (see Havas 1989 and Kennefick 2005).

The problem of motion in general relativity is essentially the same problem studied as celestial mechanics throughout the eighteenth and ninteenth centuries from Newton to Poincaré. The chief difference is that, in general relativity, unlike Newtonian theory, even the two-body problem for equal-sized bodies cannot be solved analytically. The subject got off to an excellent start, however, with the discovery by Karl Schwarzschild (1916) and others of an exact solution of Einstein's equations describing the gravitational field of one mass, whether a singularity or an extended body. This breakthrough has allowed perturbation theory to be used to describe the motion of a small particle in orbit around such a body. An alternative approach, in the case of bodies of equal size, is to treat them as weakly interacting, slowly moving masses, thus reducing their motion to that of Newtonian theory, and then to calculate corrections to that motion based on general relativity, expressing the solution in powers of such small parameters as the system velocities (relative to the speed of light, v/c) and field strength (GM/rc^2). This expansion scheme became known as post-Newtonian and was developed in the early days of general relativity theory by Droste (1917) and de Sitter (1916). Both of these approaches led to confirmations of Einstein's early result that found the missing contribution to Mercury's perihelion shift, the single greatest contribution of general relativity to classical celestial mechanics to date. In his 1916 paper that also discussed gravitational waves, Einstein himself introduced the linearized approximation, which later formed the basis for a fast-motion (or postlinear or post-Minkowski) expansion of the equations of motion, in which the expansion parameter was the field strength, with the velocities unrestricted, so that the bodies could move very rapidly.

In short, the two basic approximation schemes used in general relativity were based on the two available comparisons with existing theory: on the one hand, the previous Newtonian gravitational theory; on the other hand, the special relativistic theory of the electromagnetic field. However, for the theory of gravitational waves, each of these presented a fundamental problem. Newtonian gravity had never allowed for the possibility of gravitational radiation. Suitable as it was for supplying corrections to the equations of celestial motion for two bodies, the post-Newtonian expansion proved to be both ambiguous and ad hoc in its depiction of radiation effects. The postlinear, or fast-motion expansions, on the other hand, while eminently suited to describe radiation, were difficult to handle when applied to the problem of motion of gravitationally bound objects. The linearized theory was not strictly speaking a gravitational theory, lacking as it did the characteristically nonlinear features of the gravitational force. In fact, the basic linear metric is quite flat, with no curvature at all. It is actually the metric of special relativity theory, named for Hermann Minkowski, who first formulated special relativity theory in an invariant four-dimensional way.

A thorough account of the problem of motion before the Second World War is given by Havas (1989), who concluded that in the postwar period almost all the prewar work was forgotten or ignored, with the exception of the Einstein-Infeld-Hoffmann work mentioned above. Two important reasons for that neglect are to be found in the extreme dislocations caused by the war, the death and exile of some of the participants, and the failure of some early work to be translated into English until long after it had been written. The fact that Britain and, especially, America were important refuges for displaced scientists during the war is obviously greatly responsible for the important shift in the lingua franca of physics that occurred at this time.

Einstein's own prestige must have contributed to the relative prominence of the EIH method, and we can, with Havas, also assign some credit to the successful promotion of the scheme by Infeld, who was a prominent figure in the life of the postwar relativity community. In any case, today's textbooks still cite EIH as the canonical solution of the problem of motion. Among other factors contributing to the eclipse of prewar work was the unfashionable status of general relativity immediately before and after the war, which discouraged work in the field; the fact that much of the work predating EIH suffered from minor calculational errors, so that they did not correctly derive the EIH solution; and the fact that Vladimir Fock, who did continue to do important work on the problem after the war, was viewed with suspicion by most relativists because of his unorthodox views on general covariance.

On the other hand, despite the prominence of EIH as *the* solution of the problem of motion, James Anderson (1995) has insisted that most subsequent attempts to extend the problem of motion, especially in the direction of radiation reaction (not dealt with in any of the prewar work on the problem of motion), failed to appreciate or take advantage of the best points of the EIH scheme. In his view, EIH, which he counts among Einstein's most significant work, was the great lost scheme of the postwar period, and the back-reaction problem suffered by a tendency to ignore this "new approximation method," as Infeld called it (Infeld and Wallace

1940, p. 806). The extent to which EIH has received a very mixed press is indicated by the marked disagreement between Havas and Anderson, the former decrying its malign influence on the field, the latter lamenting its lack of influence. Havas was and is a trenchant critic of EIH, whereas Anderson now regards it as the most significant work on the problem of motion in general relativity. Both men are talented researchers with a strong interest in the history of their own field. If they can disagree so wildly, we see that we have to proceed carefully.

In the 1940s and '50s, when attempts were first made to extend the problem of motion to the order of approximation at which back-reaction effects would appear, there were several points at which EIH was felt to be wanting as a tool.[1] One was its use of point sources. Einstein himself was troubled by the use of singularities to represent the masses in motion, in spite of the long history of point-source calculations in celestial mechanics dating back to Newton himself. Accordingly in developing EIH, he made use of a clever device in which all calculations to do with the point masses were taken by integrating over a surface surrounding the point or singularity. Thus, all that mattered for the field outside the surface were the quantities calculated on the surface, which ought to be the same (to first approximation) no matter what was inside, whether a suspicious singularity or a complexly modeled planet or star.[2] It was this trick that excited the admiration of Anderson at a later date, but the use of singularities greatly troubled many relativists, even if the points had been hidden away behind a concealing surface. As we shall see, Bondi was of the opinion that radiation of gravitational waves from binary stars depended on the internal forces at play within the stars and not on the gravitational forces between them, the only kind dealt with by EIH. Bondi's concerns motivated the celebrated Indian astrophysicist Subramanian Chandrasekhar to do his own "slow-motion" calculation of the post-Newtonian type with reasonable stellar models.

Already in the 1940s, efforts were made to improve upon EIH by dispensing with the point sources. While working at the Princeton Institute for Advanced Studies, Hu Ning (known as Ning Hu in the West, with surname last), a Chinese physicist and a graduate of Caltech, was encouraged by Wolfgang Pauli, one of the few mainstream physicists then interested in general relativity, to apply EIH to the problem of gravitational radiation. Hu then went to the Dublin Institute of Advanced Studies, then best known for the presence of Erwin Schrödinger. In his pioneering work, he attempted to adapt the basic EIH scheme with the use of extended sources. He presented his results to the Royal Irish Academy in Dublin, reporting an energy loss disagreeing with the quadrupole formula in the case of an equal-mass binary system in a circular orbit (Hu 1947). Shortly before publication, however, he added a note in proof after finding a calculational error that changed the sign of his result, giving antidamping instead of damping. In other words, the system would gain, rather than lose, energy as the result of emitting radiation. The binary would therefore slowly increase in radius, not decrease.

Hu returned to the problem of gravitational waves decades later when he was a professor at Beijing University. In the late '70s the quadrupole formula controversy, which concerned the validity of Einstein's formula for wave emission, was raging fiercely, and Hu revisited his old work. By 1979 he had concluded that the fault in his original calculation from the forties lay in a failure to impose the correct boundary conditions (Hu 1982). This was perhaps the most bedeviling problem encountered in the slow-motion approach to the binary star radiation problem. The problem is that the behavior of the source system cannot be separated from its interaction with the rest of the universe. In theoretical problems involving isolated systems, the rest of the universe is actually represented by the gravitational waves themselves as they travel very far from the system, towards (or from) infinity. Suppose we imagine that the rest of the universe is sending gravitational waves *inward* to the system. In that case it seems logical that the binary system will gain energy and the stars will spiral outward, moving away from each other. More mechanical energy in the system means the two stars are closer to escaping from each other altogether; with less energy they come nearer to falling into each other.

A more physically plausible scenario involves gravitational waves radiating *outward* from the source, carrying energy away from the system, which therefore causes the stars to spiral towards each other. If as much gravitational wave energy is coming inward as going outward, then the stars should remain at the same distance from each other, in equilibrium. This is the scenario discussed as a possibility by Einstein in his paper with Rosen.

Experience, what we may call the analogy with electromagnetism, suggests that sources lose energy as they radiate. This indeed is one important aspect of what is known as the arrow of time. We know that in the future a source of radiation will have distributed its energy to a very large number of absorbers. What will not happen is that all of those absorbers out there, themselves also sources, will spontaneously conspire to radiate together so as to give our source more energy than it started with, at their collective expense. This is one physical way of looking at the problem. Unfortunately, as Hu discovered, and did others, it is not at all transparent how one applies this correct scenario to the problem. To do so, one must work out the correct boundary condition to be applied to the equations to represent just this case. Because the slow-motion problem is focused on the source and does not have much mathematical apparatus for describing the infinite distances far from the source, it is very easy to choose boundary conditions very different from what one intended. If these "unconscious" boundary conditions describe, however inadvertently, an influx of radiation from infinity, then the calculation will show the source gaining energy and spiraling outward. In 1947 Hu proposed that this mysterious behavior might be associated with the then relatively new idea of the expansion of the universe. By 1979 he had concluded that he had made an error in his boundary conditions. Essentially he had inadvertently introduced into his calculation some fictitious gravitational waves beaming into his binary star system from the rest of the universe. Since there seems to be no reason for the rest of the universe to behave in this fashion in real life, he eventually concluded that his calculation had been in error.

This whole issue is another way of talking about the problem of retarded and advanced potentials, already addressed in the discussion of Einstein's paper with Rosen in chapter 5. The boundary conditions involving waves radiating outwards from the source are linked to retarded potentials, in which the motions of the source are felt by the distant receivers at a later time. The waves converging inwards upon the source represent a boundary condition linked to advanced potentials, in which the distant receivers get news of the motions of the source before they have actually occurred. Looked at this way, it is easy to see that the converging waves are simply the diverging, physically realistic waves, observed moving backwards in time. If one makes a movie of ripples on a pond radiating outward from a source, say a stone thrown into the pond, and then plays the movie backwards, it will appear as if waves are being sent out by the shoreline and converging inwards on a certain point, out of which the stone will miraculously appear at the end of the film. Thus in choosing our boundary conditions (or our advanced versus retarded potential), we are simply choosing the direction of the arrow of time. If one chooses the reverse time direction (even without intending to), then obviously the binary star, instead of spiraling inward, will spiral outward.

Now no one would seriously propose that gravity works in a time-reversed way compared to everything else in the universe, but what about the possibility of half-advanced-plus-half-retarded potentials? In this case one is supposing that there will be an equal mix of outwardly moving waves and inwardly converging ones, so that there is no net flow of energy between the source and the rest of the universe. The waves will appear to be standing waves, frozen in time. Again, this is the scenario Einstein discussed in his paper with Rosen. What one has to keep in mind is that many calculations in field theory actually make use of half-advanced-plus-half-retarded potentials. Indeed, EIH is a good example of this. The reason is that this kind of potential allows one to assume that a change in the field of one moving particle is felt instantaneously by all other particles throughout space. This assumption can be a considerable simplification in calculating motion problems. It is essentially a return to the action at a distance of the old Newtonian gravitational theory and works for problems where radiation is not involved, which is true of the EIH paper, which tried to deal with situations in our solar system, where gravitational radiation plays no role. There are two reasons to be doubtful about the use of this potential. One is a highly technical objection that in the full theory of general relativity the equations are nonlinear, unlike those of electromagnetism. It is thus not guaranteed that if the retarded and advanced potentials are each a solution of the equations that their linear combination must also be a solution. However, when dealing with schemes which approximate to theories where this maneuver is admissible (Newtonian gravity, electromagnetism with special relativity), the choice seems quite natural. The second reason for concern would be that the choice of this potential seems to presume in advance that there will be no loss of energy due to radiation. But as we have seen in chapter 4, there was, in the early days at least, almost a hope amongst some physicists that orbiting systems would not decay due to radiation.

There had been those, such as Ritz and Tetrode, who had proposed that the equations of electromagnetism, which allow time-reversed solutions as easily as the physical solutions we observe, should best be analyzed using the action-at-a-distance approach. The most celebrated instance of this approach, and the one which really focused attention on the role of radiation in the problem of the arrow of time, is the Wheeler-Feynman absorber theory. In 1908 Einstein and Ritz had debated the role played by time in electrodynamics. What lay behind the fact that radiation processes are "irreversible," like those processes in thermodynamics which are characterised by an increase in entropy? According to the joint statement issued by Einstein and Ritz concerning their debate.

> Ritz considers the restriction to the form of retarded potentials [in electrodynamics] as one of the roots of the second law [of thermodynamics, [i.e., the law of entropy], while Einstein believes that irreversibility is exclusively due to reasons of probability. (Einstein and Ritz 1908, p. 324)

In other words, Ritz believed that the explanation for the law of increasing entropy in closed systems is that electrodynamics is restricted to the retarded potential only. In contrast, Einstein believed that that the equations of electromagnetism are indifferent to the arrow of time, just as other fundamental laws of physics are, and as in the case of thermodynamics, the arrow of time only arises when one considers the probable behavior of systems consisting of enormous aggregations of particles. It was Wheeler and Feynman who eventually took up this challenge and showed that one can begin by assuming that all electromagnetic sources interact instantaneously (in which case they respond partly to events in the past and partly to events in the future) and end up discovering that there is an arrow of time, as long as the total number of absorbers is enormous. The reason is that when a source emits to many absorbers, they are stimulated to make their own emissions, which act back upon the original source. They become sources in their turn. Wheeler and Feynman's calculations showed how those emissions from these stimulated sources, reaching the original source with no time delay, behave so as to cancel out the advanced part of the original source's potential. Thus only the retarded part of the potential survives, and the source behaves as if there is an arrow of time. As Wheeler described it in his autobiography:

> The startling conclusion Dick Feynman and I reached, which I still believe to be correct, is that if there were only a few chunks of matter

in the universe—say only the Earth and the Sun, or a limited number of other planets and stars—the future would, indeed, in reality, affect the past. . . .

When working on our second action-at-a-distance paper (published in 1949) Feynman and I went to call on Einstein and see what he had to say. "Well," he said (I am paraphrasing), "I have always believed that electrodynamics is completely symmetric between events running forward and events running backward in time. There is nothing fundamental in the laws that makes things run only in one direction. The one-way flow of events that is observed is of statistical origin. It comes about because of the large number of particles in the universe that can interact with one another." Once again, Einstein exhibited his astounding intuition about the physical world. Without having done any of the calculations or analyses that Feynman and I had been through, he had reached the same general conclusion about the role of distant absorbers in influencing what happens here and now in small spaces and times. (Wheeler 1998, pp. 166–167)

The question of the origins of the arrow of time is a compelling one for physicists, even if it is too subtle to figure prominently in their day-to-day research. In 1963, Thomas Gold organized a conference on this topic (supported by the USAF), to which Wheeler and Feynman were invited as leading experts. When the deliberations of the meeting's participants were published in a book (Gold 1967), Feynman objected that his remarks had not been intended for publication, and so his contributions were attributed in the book to a mysterious Mr. X. As mentioned earlier, Rosen made an effort in 1979 to argue that what Feynman and Wheeler had shown for electromagnetism would not work in gravitation, because gravity is notoriously weaker than other forces. Rosen argued that the back action of the absorbers on the source would simply not be strong enough to cancel out the advanced part of the potential. Although this argument did not provoke much debate, it did reflect the longstanding skepticism of whether gravitating systems in orbit around each other really were irreversible systems that obeyed the arrow of time.

This idea, which relativists seemed hardly to dare to hope for, that the Earth's motion around the Sun is somehow exempt from the arrow of time, of course reflects deep emotions and old debates. Just as Newton is said to have seen the inevitable decay of the solar system as a sign of God's plan for the end of days, many other people have seen the eternal clockwork universe of Laplace as the only kind of immortality of a fragile world. Can it really be true that decay has made its way back into the perfect system of celestial mechanics after all? One is reminded here of the old couplet of Alexander Pope's, completed by John Collings Squire:

Nature and Nature's laws lay hid in night: God said, "Let Newton be!" and all was light.

It did not last: the devil howling "Ho! Let Einstein be!" restored the status quo.

In addition there were technical reasons that might encourage relativists to hope for a reprieve from the temporal world. A static, time-independent solution to the problem of binary motion would permit easier and more accurate relativistic solutions to the problem of motion of stars and planets. As a matter of fact, even in electromagnetism there had been efforts to use action-at-a-distance field potentials as a way of exempting certain systems from radiating. In the early twentieth century it was a complete mystery why electrons orbiting the atomic nucleus did not undergo a continuous decay in their orbits as they radiated away energy in wave form. Before this phenomenon was explained by quantum mechanics, physicists like Tetrode looked for a classical solution by appealing to the half-advanced-plus-half-retarded potential (others who did so included Gunnar Nordström, discussed earlier, and the American physicist Leigh Page [Page 1924]). Wheeler said in his autobiography that he was attracted to action at a distance because of his efforts to construct stable models of atomic particles out of smaller fundamental particles, systems that would need to orbit but remain static (Wheeler 1998, p. 12).

Certainly one person who reported to his own students in later years that Einstein had hoped for such an outcome in the case of binary stars was Leopold Infeld (Andrzej Trautman, interview). In the years after World War II he continued to investigate the problem of motion, at first with Einstein and then with his own students.

Not long after the EIH paper was successfully completed, Infeld, who had with Robertson's help secured a position at the University of Toronto, put his graduate student Phillip Wallace to work applying the EIH formalism to the problem of motion in electrodynamics. In their paper, as in the

EIH paper itself (where radiation effects were not considered), we see a preference for the standing wave potential, half-advanced-plus-half-retarded. Infeld and Wallace state that this solution "does not specify a privileged direction for the flow of time" and is, besides, the simplest for their method (Infeld and Wallace 1940, p. 799). They note that this solution does not damp orbital motion and further state that "the addition of radiation seems from this point of view arbitrary," since one must choose the retarded potential to obtain it. This viewpoint may partly reflect Einstein's own, but it should be stressed that he was the first to make use of a retarded potential in general relativity in his seminal 1916 paper on gravitational waves. The solutions that admit radiation damping are objectionable because they involve an arbitrary imposition of the arrow of time into field theories that are otherwise time symmetric. Einstein had thought that time asymmetry had no business in field theories and that its origins lay solely in probability theory. His views may have influenced Infeld, who preferred the standing-wave solution as the most natural choice in the EIH approximation. In the case of the gravitational field, where the existence of radiation could not be experimentally proven, Infeld may have felt there was no compulsion to impose the arrow of time, as one would in electromagnetism, knowing from experiment that radiation existed in that field.

Subsequently Infeld and another student, Adrian Scheidegger, worked on the problem of gravitational radiation reaction in the EIH formalism (Infeld and Scheidegger 1951). They concluded that the most natural treatment of the scheme, employing the half-advanced-plus-half-retarded potential, led to a no-radiation-reaction result. It was possible, they conceded, to find terms at certain large, odd powers of v/c (where c is the speed of light, and v represents the small source velocities) that appeared to correspond to back-reaction terms, but they contended that these could always be transformed away by a suitable choice of coordinates. The result, when announced at an American Physical Society meeting in 1950, "gave rise to a considerable flow of discussion," as Scheidegger put it (Scheidegger 1951). That same year Infeld left Canada, after a McCarthyite campaign against him in the press (see chapter 9). He returned to his native Poland, while Scheidegger continued to argue the no-damping position in North America in his absence, before leaving the field of general relativity for that of geophysics in the mid-fifties because of the shortage of available academic positions.

How could Infeld and Scheidegger seriously suggest that their choice of potential, which seems to preempt the possibility that binary stars emit gravitational waves, could be physically the correct one? Doesn't this fly in the face of our experience with the arrow of time? Of course, the fact that there are many irreversible processes in nature does not mean there are no reversible processes, which do not decay. Experience with electromagnetism suggests that charges which are in accelerated motion must radiate electromagnetic waves, and the quadrupole formula suggests the same for gravity. Stars in orbit around each other are certainly accelerating, and the quadrupole formula, if applicable to them, would indicate that they lose energy to radiation. But the very fact that the gravitational field equations completely determine the motion of a system means that one cannot rule out the possibility that the motion which the field equations constrain the stars to follow just happens to be the very same motion which would cause them not to lose energy to wave emission. In electromagnetism we are free to say, the bodies move in such a way, which results in the following radiation being emitted. But in the gravitational problem of motion we cannot do this. The bodies move in a way that satisfies the field equations, which themselves take on a form that depends on the motion of the body. It appears as if we must know beforehand how the bodies are moving in order to figure out how they move! The solution is to use iteration. Begin by imagining that the bodies do not radiate at all. Now we know that they follow geodesics of the spacetime (the shortest route between two points, which is not a straight line in the curved spacetime of relativity). So assume this and work out what radiation they will emit as a result. But this radiation will carry energy away, and we derived the motion by assuming no energy was lost. So now go back to the motion and modify it to allow for the lost energy, so that the stars no longer follow geodesics but some other path, probably not very different. Now calculate the resulting radiation from that path, and then reapply that to the motion, altering it again. Repeat the process as many times as desired to achieve an accurate result, according to your best judgment.

Therefore, in the special case of a system held together entirely by the force of gravity, we can never know for sure that the motion which would result if we took radiation into account might not be the very motion which resulted in no net energy loss to radiation at all. As usual, the skeptics emphasized this particular way in which general relativity appeared to differ from electromagnetism. As Scheidegger observed, relativity occupied a "peculiar place" among classical field theories (Scheidegger 1953). The important peculiarity is that the equations of motion are constrained by the field equations, as Einstein had noted. In electrodynamics, where this is not the case, we are perfectly free to demonstrate damping effects by moving the particles around in whatever fashion and showing that this gives rise, when the field equations are invoked, to radiation and loss of energy from the local system. In relativity, it is necessary to show that the motions in question are permitted by the same field equations.

This is all the more important when we consider the question of what *type* of motion gives rise to radiation. One obvious example is an accelerating charge in electrodynamics. The issue here was nicely expressed by Feynman in his 1962–63 lectures on gravitation at Caltech. In the same course, incidentally, Feynman characteristically dissented from the common opinion among relativists about relativity's "peculiar" field equations, arguing that relativity, in this respect, was no different than the electromagnetic field theory. About the connection between the equivalence principle and electromagnetic radiation, he had this to say:

> The principle of equivalence postulates that an acceleration shall be indistinguishable from gravity by any experiment whatsoever. In particular, it cannot be distinguished by observing electromagnetic radiation. There is evidently some trouble here, since we have inherited the prejudice that an accelerating charge should radiate, whereas we do not expect a charge lying in a gravitational field to radiate. This is, however, not due to a mistake in our statement of equivalence but to the fact that the rule of the power radiated by an accelerating charge,

$$\frac{dW}{dt} = \frac{2}{3}\frac{e^2}{c^3}a^2, \tag{8.1}$$

[*dW/dt* is the power radiated by a charge *e* moving with acceleration *a*, *c* is the speed of light.] has led us astray. . . .

Of course, in a gravitational field the electrodynamic laws of Maxwell need to be modified, just as ordinary mechanics need to be modified to satisfy the principle of relativity. After all, the Maxwell equations predict that light should travel in a straight line—and it is found to fall towards a star. Clearly, some interaction between gravity and electrodynamics must be included in a better statement of the laws of electricity, to make them consistent with the principle of equivalence.

We shall not have completed our theory of gravitation until we have discussed these modifications of electrodynamics, and also the mechanisms of emission, reception, and absorption of gravitational waves. (Feymann 1995, p. 123)

Since Feynman had raised the question but not answered it in the portions of his course which have been preserved, what of the case of the radiation (no matter whether it is electromagnetic or gravitational radiation) emitted by a falling mass? The falling object is clearly accelerating with respect to the person who dropped it, but in a relativistic sense it is merely following a geodesic and is thus entirely unaccelerated. To put it in a manner that would make sense to Aristotle, it is merely doing what comes naturally. In terms of the local spacetime, the particle that is really being *accelerated* would be the one still being held in or supported by the observer's other hand, which is prevented from falling freely. The observer's hand is accelerating the particle to prevent it from falling (following a geodesic). For instance, a person in free fall would believe the particle in free fall was stationary and the supported particle was accelerating. Though we can readily appreciate that acceleration is purely in the eye of the beholder, surely the same is not true of electromagnetic radiation? Which one of these particles *ought* to radiate? Ezra T. Newman, one of the leading young relativists of the period following the Chapel Hill conference, related how Wheeler once asked a roomful of relativists to vote on this question, and Newman recalled the room being fairly equally divided (interview). Unfortunately, in the context of relativity theory, it is not trivial to identify what acceleration in the radiative sense is, even for the experts! This seems to be a rare example of the democratic approach to science.

Nowadays relativists agree that the question of which particle emits radiation depends on the observer's state of motion. Someone falling freely will think it is the supported particle emitting radiation; someone not falling, but just standing there in the laboratory, will think it is the falling particle which radiates. A distant observer in some other part of the universe will agree with the person who is not in free fall. To them the falling particle and the Earth are accelerating towards each other, and this motion produces a changing field that is related to an electromagnetic or gravitational wave which is observable far from the source. But why then does the free-falling observer fail to see waves emitted by the free-falling particle? Why does she, even more paradoxically, see waves being emitted by the supported particle, when a distant observer sees no waves at all? The answer lies in the mysteries of the near-zone field. Close to a dynamical source of any field (where *close* means "not many wavelengths away from the source," so the size of the near zone depends on the characteristic length scale of the wavelength of any radiation involved), it is not possible to really distinguish wave phenomena from other dynamical fluctuations of the field. We on Earth, for instance, feel a constantly changing tidal force exerted on us by the Moon. This force produces an effect which is indistinguishable from that produced by gravitational waves. Someone on a nearby star cannot feel the tidal force exerted by the Moon, because it is too weak. But if the Moon produced appreciable gravitational waves, they might be able to detect those. Up close you cannot tell the difference between gravitational waves and oscillating tidal forces. Far away you can do so, principally because the tidal forces are too weak to detect at great ranges. The freely falling observer and the laboratory observer fail to agree on which parts of the field are really waves. The distant observer has no such handicap.

But what about the problem of radiation reaction? If one particle is *really* radiating, and this radiation is really being detected by a distant observer, it must mean that energy is being transferred from the particle to the distant observer. In consequence the particle loses energy and this must manifest itself as a loss in kinetic energy, which means that the particle falls less slowly. There is a radiation reaction "force" which opposes the force of gravity and makes it fall less slowly. The observers can surely agree about whether a charged particle falls more slowly than an uncharged one, because the first emits electromagnetic waves as it falls and the second does not. This is probably a question about which relativists will still argue for some time to come. Some say that the equivalence principle really is violated here, the charged particle *does* fall more slowly, and therefore the equivalence principle does not apply to charged particles. Justification for this viewpoint is found in the fact that the equivalence principle applies only in a very local sense. It is easy to tell whether you are on an accelerating spaceship or inside a room on Earth if, for instance, you are permitted to open the window and observe or do experiments outside the small room. Indeed, if the room is large enough, experiments inside it without benefit of a window will suffice (particles on a spaceship fall out the back along parallel lines; particles on the earth converge as they fall towards the center of the Earth). A charged particle comes with an electromagnetic field attached. If this field is interacting with distant objects throughout the universe, the experiment seems highly nonlocal indeed.

Where the experts differ, the rest of us are well advised to put the question aside. But in this book we cannot duck the question as it relates to gravitational, rather than electromagnetic, radiation. Does the freely falling particle fall less quickly than it otherwise would have if we allow for the effects of radiation reaction? Is there an upwardly directed radiation reaction force that opposes the usual downward part of the gravitational force? The answer appears to be yes, but in this case there is no difficulty with the equivalence principle, since all particles have mass, and although the radiation reaction force will be bigger for larger masses, the amount of acceleration produced will be the same for any mass. What of the freely falling observer? She sees no radiation coming from the freely falling particle. How can she therefore explain the radiation reaction force? The answer is that she still sees this force, which is the consequence of the self-interaction of the particle with its own retarded field (that is to say, its field emitted at an earlier time and scattering back to the particle off the curved spacetime of the planet). Her view is like that of Laplace in his eighteenth-century calculation: it takes into account the finite speed of propagation of the gravitational field but is blind to the radiation associated with it.

In his talks at this time, including the one at the Chapel Hill meeting, Bondi pointed out the distinction between two masses being waved about at the end of someone's arms, clearly not following geodesics and clearly emitting gravitational waves (but tremendously weak ones!), and two masses in a binary star system, following geodesics and, if the skeptics were right, not radiating anything (DeWitt 1957, p. 33). Since gravitational forces were likely to be the only forces capable of moving large masses very quickly, the issue of whether purely gravitational systems could give rise to radiation was an issue of whether such radiation would ever be detectable. To the surprise of most theorists, that issue of detectability, would, thanks to Weber, become one of some practical interest. But in the meantime the theorists were determined to address the problem nonetheless.

A direct response to Infeld and Scheidegger came in 1955. Joshua Goldberg, whom we already met in his role with the air force, was a student of Peter Bergmann's, who had criticized the Infeld and Scheidegger results. Bergmann, another German-speaking refugee from the Nazis, had been one of Einstein's assistants at the same time as Infeld and Hoffman. After the war he mentored a generation of students, founding perhaps the leading, and certainly the earliest, school of relativity of the postwar period. Goldberg examined the reaction problem in the EIH formalism (Goldberg 1955), and his conclusions were twofold. On the one hand, he denied that the slow-motion approach tended to exclude the possibility of damping, arguing that coordinate transformations which removed some back-reaction terms would reintroduce other reaction terms of odd order in v/c. In particular he showed how some terms associated with radiation (defined in a way which "agrees with that used in the theory of electromagnetism" [p.1873]) contributed to the Riemann curvature tensor and therefore could not be removed by coordinate transformations. On the other hand, he determined that it was poorly suited to the back-reaction problem, principally because of the restriction to slow motions of the source. In fact, it was generally agreed that radiation reaction terms did not enter into the post-Newtonian equations of motion until terms of order at least $(v/c)^5$ beyond Newtonian order (or post-$2\frac{1}{2}$-Newtonian order). Since first-post-Newtonian effects (or $(v/c)^2$ order), such as those obtained by EIH, were both small and difficult to calculate, the expansion method seemed unpromising for studying radiation in that it had to be pushed to high order to succeed.

A couple of years later Goldberg was introduced to Peter Havas, a physicist with experience in the problem of radiation in electrodynamics who shared Goldberg's interest in developing a fast-motion expansion in general relativity. Havas had been a student of Guido Beck's, although not in Vienna, where they both began their scientific careers, but in Lyons, France, where they both sought refuge after the Anschluss in 1938. Beck had worked in the Soviet Union for a period, in Odessa, Ukraine, at the same time Rosen was in Kiev but had to leave when foreigners began to be arrested for espionage. He then found himself rendered stateless by the Nazi invasion of Austria. Havas, whose studies had been interrupted by the invasion, had worked as an experimentalist in Vienna under J. Mattauch and was continuing his studies at the Atomic Physics Institute in Lyons when he encountered Beck there. Seizing a chance to move into theory, Havas persuaded Beck to supervise him. This work was in the mainstream subjects of atomic and nuclear physics, as Beck no longer worked in general relativity.

Following the outbreak of war, Beck and Havas were interned by the French government, as was Bondi by the British, as we shall see. Initially set to forced labor in the camps, Havas was fortunate to be rescued by his future wife (also Austrian; she had not been interned because only males were), who managed to effect his release after a few months, and he returned to experimental work at Lyons. The city was briefly occupied by the Germans after the invasion of France, but Havas and his wife made good their escape before the former's arrival, returning when the city was handed over to the control of the Vichy collaborationist government. Since Beck was still interned, Havas continued his theoretical studies by correspondence while working with the nuclear fission group at Lyons. In late 1940 Beck was released, and the next year Havas managed to secure a visa for the United States, where he finally completed his studies at Columbia University in New York. Beck eventually escaped to South America via Portugal (Havas 1995).

In the United States, Havas secured a position at a college in Pennsylvania, Lehigh University, and worked on electrodynamic theory. His experience in this field convinced him to try something similar within general relativity (inspired by the discussion of the EIH work in Infeld's autobiography, *Quest*), and he contacted Goldberg to secure an invitation to the Chapel Hill meeting in 1957 (see above), where he encountered a considerable level of interest in the radiation problem among relativists. He was particularly inspired by Bondi's talk addressing the differences between electromagnetism and gravitation in the radiation problem (Havas, interview). Being familiar with the special relativistic problem of radiation, his instinct was to approach the problem via what would later be called a post-Minkowski approximation, that is, fast motion, which essentially approximates general relativity to the special theory rather than to Newtonian theory. Since Goldberg had independently reached a similar conclusion—that this was a more appropriate way to handle the back-reaction problem—they began a collaboration based on this approach.

Infeld's move to Poland acted to quieten the controversy over radiation damping of binary stars in the west, but the debate simply moved with him to the East, where Infeld conducted a sharp rhetorical struggle with the Russian physicist, Vladimir Fock, a very prominent theoretician known for his work on quantum mechanics. Fock had developed a slow-motion expansion independently of EIH in the late thirties (Fock's original paper had appeared in Russian in 1939 but was little known in the West). In 1955 he published a book, the English version of which appeared in 1959 under the title *Spacetime and Gravitation*. In it he included his work on the problem of motion and radiation damping, for which he made use of outgoing-wave-only boundary conditions. His results were in agreement with those of Landau and Lifshitz, which is to say that he found that the leading-order radiation damping of the orbit obeyed the quadrupole formula prediction. Perhaps the most noteworthy aspect of Fock's method was his use of matching techniques in imposing boundary conditions on the problem of motion.

Because the slow-motion approximation scheme reduces elements of the theory to the old Newtonian theory, the approximation scheme does not describe the full four-dimensional complexity of spacetime far from the source. But it is precisely in this region very far from the source that one has to enforce the boundary conditions. Close to a physical source of gravitational waves there are many other gravitational effects, many of which are bigger than the effect of the waves. For instance, tidal effects, which feel very much like gravitational wave effects, are very strong near a source. Think of the way the Moon raises tides on the Earth, yet the amount of gravitational radiation emitted by the Earth-Moon system is completely undetectable at any distance from the source. However, tidal and other gravitational effects die off very quickly as you move far from the source. It is well known that the force of gravity falls off as the inverse square of the distance from the source. Tidal forces (which depend on the gradient of the gravitational field), by contrast, fall off as the inverse *cube* of the distance from the source, which is to say, even faster than the simple attractive force. This is why the Earth feels the gravitational pull of the Sun more than that of the Moon, but the tides raised in our seas by the Moon are much bigger than those raised by the Sun. The Moon is smaller than the Sun and so pulls on us more weakly, but it is closer; the Sun's tidal influence falls off with distance so rapidly as to make it less significant. The size of a gravitational wave, on the other hand, falls off at only the inverse of the distance from the source.[3] So gravitational sources in other galaxies have almost no chance of noticeably influencing Earth by their attractive pull or their tidal forces because of the incredible distance involved, yet we can still hope and even plan to detect gravitational waves from these places.

As mentioned already, close to the source, in what is called the near zone in gravitational wave terminology, one cannot even hope to distinguish between the waves and the other effects, such as tidal effects. It is only in the distant zone, the so-called wave zone, that you can even hope to identify a gravitational wave distinct from the other gravitational effects of a source. So the boundary conditions were imposed by Fock on a rather different wavelike solution in the wave zone, towards infinity. Fock then "matched" his two solutions in an intermediate region to impose the boundary conditions on the equations of motion for the system. It was not readily apparent in the EIH scheme how those conditions were to be imposed on the solution of the sources motion so as to match correctly the source with its radiation field. In this case, Fock's prejudices proved more useful than Infeld's. Fock was quite clear on the reasons for the superiority of his method, declaring in a section titled "On the Uniqueness of the Harmonic Coordinate System":

> When solving Einstein's equations for an isolated system of masses we used harmonic coordinates and in this way obtained a perfectly unambiguous solution. We found unique results not only for finite and "moderately large" distances from the masses, when the wave-like, i.e. hyperbolic [that is to say, finite propagation speed], character of Einstein's equations was not essential and was accounted for by the introduction of retardation conditions, but also for the "wave zone."(Fock 1959, p. 346)

After his return to his native Poland in the fifties, Infeld argued with Fock that general covariance was an essential feature of Einstein's theory and the fact that Fock's radiation reaction result appeared to rest on this special choice of coordinates no doubt encouraged him in dismissing it. He could not accept that the harmonic gauge condition was the only suitable choice for radiation theory, since Fock claimed that this suitability arose from an underlying correctness in this choice of coordinates. For Fock, the fact that the harmonic choice of coordinates led to "a much simpler" calculation arose from the fact that

> [h]armonic coordinates are the nearest in their properties to ordinary rectangular [i.e., Cartesian] coordinates and ordinary time in the

Minkowski "world" [i.e., the spacetime of special relativity]. That is why, in these coordinates, the GTR formulas are the clearest. (Translated and quoted in Gorelik 1993, p. 315).[4]

Even after his book was translated into English, it wielded little influence in the West, not only because of his opposition to general covariance, but because his philosophical views were couched in the rhetoric of Marxism; At least from the thirties on, Fock had been strongly influenced by Lenin's writings on science. Lest Fock be suspected of opportunism in adopting this philosophical line, I should remark that he was instrumental in preventing general relativity from being anathematized during the period of strong Soviet reaction against Einsteinian "idealism." He showed extraordinary physical and moral courage during and after the purges by supporting scientists who had been arrested and by continuing to give them credit for their discoveries when others feared even to name them in public (Gorelik 1993, pp. 312–313). In respect of their defense of relativity against Soviet "idealists," Fock and Infeld fought the same fight. They disagreed over general covariance, an argument that has gone Infeld's way since, and also over gravitational radiation reaction, where it is Fock whose views have been vindicated by history. Even at the time, in fact, Fock's "Marxist" views on general covariance, were not accepted by the leading Soviet theorists, as Infeld himself observed (Gorelik 1993, p. 320; see also Graham 1987). On the other hand, there were, in the Soviet Union, no important skeptics of the existence of gravitational waves, certainly at this period.

    Another reason why Fock's results were not so influential, apart from his rejection of general covariance and his Marxist learnings, is perhaps that Fock himself regarded his back-reaction result as merely demonstrating that wave phenomena played an inconsequential role in the problem of motion in gravity, due to the small size of the effect for known astronomical systems. In this he followed Landau and Lifshitz, who did derive the quadrupole formula, but stressed its irrelevance for astrophysics.

    In the late fifties, the EIH approximation was developed further by Andrzej Trautman, a young researcher in Infeld's group in Warsaw, who departed from Infeld's approach by adopting outgoing-wave-only boundary conditions. Trautman had been encouraged to work on the radiation problem by Jerzy Plebanski, Infeld's collaborator, who did not share Infeld's skepticism regarding the existence of gravitational waves. Colleagues of Trautman's describe how he came to work on the problem:

> His line of research [at the Institute of Theoretical Physics at Warsaw University] , namely gravitational radiation, was one suggested by [Jerzy] Plebanski, but Plebanski soon went off to Princeton on a Rockefeller Fellowship, and Andrzej "fell into the hands" of Infeld, who got him to study the radiation problem within the ill-suited Einstein-Infeld-Hoffmann (EIH) formulation. At that time, Infeld believed, perhaps because of Einstein's influence, that gravitational radiation was not emitted by freely gravitating bodies, and the EIH approach seemed to suggest this. However Andrzej's good physical instincts (no doubt heightened by his experience with radio engineering!) [Trautman had trained as an engineer before joining Infeld's physics group]—and also from his studying of an important paper by Josh Goldberg—convinced him that Infeld was wrong and that gravitational radiation [from binary systems] must indeed occur. . . . Andrzej spent many hours trying to convince Infeld that gravitational radiation must be emitted.(Penrose, Robinson, and Tafel 1997, p. A2)

Trautman confirmed Goldberg's earlier claim that the net back-reaction effect could not be transformed away but merely moved between one order in the expansion and another. Thus, his paper addressed the following question, arising out of the Infeld, Scheidegger, and Goldberg debate, "whether the situation in GR resembles that in Newtonian mechanics rather than that in electrodynamics." The possible resemblances are described as follows:

> In Newtonian mechanics the initial positions and velocities of pointmasses determine their motion completely. The situation is different in electrodynamics where the initial values of the field are required besides information concerning changes. Two *free* point-charges of opposite signs may move uniformly around a circle in a standing-wave electromagnetic field. However, the same charges may alternatively produce outgoing radiation. Their motion will not then be periodic; they will undergo damping. Which of these cases occurs in any particular system depends on the initial and boundary conditions. (Trautman 1958b, p. 627)

So the question at issue, according to Trautman, was whether or not the analogy with electromagnetism, which gives rise to gravitational radiation, should be favored over the analogy with Newtonian mechanics, in which orbital motion is not damped and radiation does not exist in the gravitational field. After going over the electromagnetic analogy and presenting a solution to the gravitational field equations in the original EIH form, he noted that

> by analogy with the scalar wave equation and Maxwell's theory, solutions of [this] form may be interpreted as representing standing-wave fields. In order to get solutions corresponding to "retarded" or "advanced" fields the series must be supplemented with the missing *radiation* terms.(p. 630)

He thus recalled the phrasing of Infeld and Wallace, where the imposition of retarded conditions was seen as arbitrary. Indeed, Trautman expressed well the underlying philosophical (as opposed to merely technical) objection to the post-Newtonian approach to the radiation problem of fast-motion advocates (like Bonnor and Havas, see below). The slow-motion expansions reduced general relativity, at first order, to a gravitational theory (Newton's) that did not admit radiation and never had. Fast-motion expansions, of course, reduced general relativity at first order to special-relativistic dynamics, a theory in which the problem of radiation was very advanced but that was not a gravitational theory. Both, therefore, had problems, which had to be overcome by pressing to higher orders in the relevant expansions, but the post-Newtonian method suffered for years from the difficulty of introducing the radiation terms in an arbitrary way. Trautman himself showed that with a correct choice of boundary condition, one could introduce radiation terms that could not be transformed away, contrary to the claims of Infeld and Scheidegger. He did not recover the quadrupole formula result, but he did, unlike Hu with his incorrect boundary conditions, find positive damping.

Specifically what Trautman showed is something that is very common to relativity theory. When one makes relativistic calculations or observations, it often appears that a measured quantity depends on the observer's coordinate system or point of view, because one is measuring or calculating only a spatial or only a temporal quantity. There is in fact an invariant quantity involved, but it has both a spatial and a temporal aspect to it.

A famous example concerns the barn paradox of special relativity, in which someone carrying a pole runs through a barn that is shorter than the pole, even though the doors were shut after he was inside the barn and opened again before he came out the other side. From the point of view of an outside observer, the problem is explained because the pole appears shortened when it is moving, according to the theory of relativity. But the man carrying the pole does not see it shortened, because relatively speaking, the pole is not moving in his coordinate system. How does he think it fits inside the barn? To him the solution is a temporal one, the two doors were never closed at the same time. Contrary to what the stationary observer sees, to the pole carrier, one end of the pole was always sticking outside of the barn, first on one side, then on the other. There is a spacetime interval, describing the total distance in both space and time between the doors closing, which all observers agree does not change. But we are used to measuring either in space or in time separately. Trautman realized that something similar was going on in EIH calculations of the orbital decay problem. The radiation terms corresponding to outgoing waves at the fifth order have a spatial character (space-space metric terms, in the relativity jargon). They can be transformed away, but only at the cost of introducing temporal terms (time-time metric terms) at a higher order in the expansion, which then introduces the same damping. The total effect represented by this combination of terms (spatial terms at fifth order in the expansion in v/c, temporal terms at seventh order) cannot be transformed away.

Whereas Fock chose a particular coordinate scheme and stuck with it, Infeld, the upholder of covariance, had experimented with different coordinate shifts in pursuit of a nondamping result. Like a good relativist, he was distrustful of anything that showed signs of noninvariance. Trautman rejected Fock's strict adherence to harmonic coordinates as "somewhat stringent," seeing "no reason for restricting ourselves to harmonic coordinates only"(Trautman 1958a, p. 409). Instead, he generalized the boundary conditions in order to investigate Infeld's old argument that the radiation terms were merely coordinate effects, successfully showing that they were not. That Infeld cannot have been convinced is shown by his remarks in his book *Motion and Relativity* (Infeld and Plebanski 1960) two years later.

Indeed, Trautman (1958a) also showed quite generally that imposing outgoing-wave boundary conditions would cause a pertain quantity associated with the flux of energy in the asymptotic waves to be zero or positive, and therefore that such conditions should result either in no radiation or in radiation carrying energy outwards from its source.

At about this time, the other longtime skeptic, Rosen, now at the Technion Institute in Israel, encouraged his graduate student Asher Peres to attack the problem as a means of deciding the dispute between Fock and Infeld over the existence of radiation reaction in the post-Newtonian problem of motion. Peres employed a method bearing some similarities to that of Fock, making use of the de Donder (harmonic) gauge condition but also employing the singularities used by the EIH method, which Rosen disliked (Peres 1959a). However, his initial results, like Hu's earlier ones gave antidamping for the binary system and thus failed to shed any light on the Fock-Infeld dispute (Peres 1959b). But after finishing his thesis, Peres realized where the problem with his previous paper lay (Peres 1959c). In imposing boundary conditions on the equations of motion for the binary system, he had inadvertently chosen conditions that included incoming as well as outgoing radiation at infinity, so that he was unwittingly introducing a source of energy to power the outward spiral of the binary.

In his paper correcting the error Peres explained the ease with which such confusion can arise:

> At each stage in the approximation procedure . . . there is a considerable freedom of choice of solutions, each representing a possible motion and a gravitational field belonging thereto. Only one of these solutions behaves at infinity as purely outgoing waves; the remaining ones also contain incoming waves. However, it is difficult to determine which solution is the correct one because the n-th term of a series expansion into powers of (v/c) behaves in the wave zone as R^{n-2} [R is the separation of the binary components], and no boundary conditions for each stage of the procedure are known. The purpose of this present note is to give a criterion which partly removes this ambiguity. . . . [The] method is not sufficient in general but it gives unambiguous results up to the seventh order. As a consequence, one has to modify the fields that were previously used and . . . [the result] agrees with [the Landau and Lifshitz quadrupole formula]. The fact that one previously obtained a negative radiated energy should be ascribed to the presence of incoming gravitational waves, which were absorbed by the particles. (Peres 1959c)

Essentially Peres's method involved identifying those terms in the expansion of the equations describing the post-Newtonian motion of the system, terms which were equivalent to those in the unretarded potential of the source. Then he would add in, "by hand" as physicists say, terms at the next order required to make those terms look like terms in the retarded potential that one would expect to find in a non-slow-motion expansion far from the source. By thus carefully disassembling and reassembling the equations of motion, one could construct a consistent outgoing-wave-only solution, by an unambiguous, if tedious, method.

Peres's 1960 paper has been referred to as containing the first correct back-reaction calculation (Thorne 1989). It was certainly an important step on the road to resolving the ambiguity plaguing this problem, an ambiguity that is nicely described by Misner, Thorne, and Wheeler in their famous textbook, *Gravitation*, written at the start of the 1970s. MTW, as this text is universally known, emphasizes that post-Newtonian schemes encounter the problem that there are no gravitational waves in Newtonian gravity:

> The dynamical part of the Newtonian [gravitational] potential, in its "standard form" . . . has no retardation in it. (Newtonian theory demands action at a distance!) Consequently, there is no way whatsoever for the standard potential to decide, at large radii [i.e., very far from the source, or "at infinity"], whether to join onto outgoing waves or onto ingoing waves. Being undecided, it takes the middle track of joining onto standing waves (half outgoing, plus half ingoing). But this is not what one wants. It turns out . . . that the join can be made to purely outgoing waves if and only if [the Newtonian gravitational potential] is augmented by a tiny "radiation-reaction" potential. (p. 993)

Nevertheless, the perceived arbitrariness of the slow-motion approach in imposing the wave-zone boundary conditions from one step in the expansion to the next, which seemed reflected in the wildly differing results produced by the method, continued to give rise to arguments that the approach was hopeless (Bonnor 1963).

The difficulty with the slow-motion expansion could be addressed by employing a different, fast-motion-like expansion in the region far from the source, in the manner of Fock. Such an expansion did admit of radiation fields, while the slow-motion approach to the problem of motion could deal successfully with the source. The following problem would then remain, how to match the two expansions in the two different regions to each other, so that whatever boundary conditions were imposed in the wave zone would be correctly applied to the solution of the source's motion? Peres was meticulous in his matching between the two regions, but as he noted himself, his method was not at all general. A completely general, unambiguous scheme would have to wait until the end of the next decade, but in the meantime, Peres and Trautman had shown a way out of the dilemma facing the slow-motion approach, even while some were pronouncing it hopeless.

While conceptually more appealing in that it took as its starting point the linear approximation in which the radiation analogy with electromagnetism had arisen in the first place, the alternative fast-motion approach, as developed by Havas and Goldberg (1962) and others (for example, Bertotti and Plebanski [1960]), was also proving frustrating. It was a difficult task to go beyond the leading-order corrections to the linearized theory, and the results of applying that step to the reaction problem, published by Smith and Havas (1965), again showed an energy gain in the source. Therefore, in his review paper of 1963, the English relativist William Bonnor concluded that the question of whether freely falling sources experienced damping remained unsettled.

Bonnor's paper contains an interesting comparison of the two main representatives of the slow- and fast-motion approaches at that time: on the one hand, EIH and EIH-like schemes, and on the other hand, the work of Havas and collaborators. With regard to EIH, Bonnor was not greatly impressed by the progress made:

> A choice of solutions of the EIH equations is available, and that made by EIH refers to the . . . non-radiative field. One can try to use the retarded potential instead, though this leads to much arbitrariness . . . Nevertheless, a number of workers have used the EIH method on radiation problems, and their conflicting results are a monument to its unsuitability for the task. (Bonnor 1963, p. 558)

Bonnor held out more hope for Havas's approach, because "unlike . . . EIH, [it] is covariant with respect to Lorentz transformations of the flat Minkowskian background metric" (i.e., it satisfies the axioms of special relativity). Nevertheless, "the linear approximation tells us nothing about radiation from a freely gravitating system" (since in the linear approximation particles are unaffected by each other's gravitational attraction), while Havas's "delicate mathematical processes" designed to overcome the "great difficulties" of the next level of approximation (first postlinear) produce an admittedly "disappointing" result, which fails to agree with the well-known perihelion shift result and, like Hu and Peres's first try, gives "an energy *gain* due to radiation, an inexplicable result." Bonnor was nonetheless still hopeful: "It would be of great interest if Havas' method could be carried one step further." With that sentiment Havas was heartily in agreement, but unfortunately that wish was never to be fulfilled.

Bonnor himself had adopted the approach of analyzing a simple mechanical system that begins and ends in a static state, a model which Bondi adopted. Like Bondi, though with a less involved method, Bonnor was able to show that such a system would lose mass, but he also concluded that "whether freely gravitating bodies [such as binary stars] radiate, and if so with what effect on the motion, is still an open question" (p. 559).

Havas and Goldberg benefited from Havas's extensive knowledge of the literature on the classical problem of motion and exploited a fast-motion approximation scheme of Havas's (1957) based on a prewar one due to the Polish theorist Myron Mathisson. Interestingly, Havas reported in a more recent paper (1989) that Mathisson was nearly chosen by Einstein as a collaborator on the problem of motion instead of his compatriot Infeld (both were Polish Jews in exile), which might have resulted in an EMH paper rather different from the EIH that actually exists. Like EIH, they employed point sources but made use of renormalization, rather than EIH's surface integrals, to avoid the resulting divergent integrals. The problem of divergent integrals (which cannot be solved by the usual methods because the result is infinite) in calculation schemes that attempted to find the effect of a point source's field on the source itself had been known for a long time in electrodynamics and had been an important topic of research since Lorentz's days. It was EIH's ability to eliminate the divergences without resorting to renormalization schemes that particularly excited the admiration of EIH enthusiast James Anderson.

As mentioned already, Havas and Goldberg's results of 1962 disagreed with the well-known result for the perihelion shift of Mercury, which the slow-motion schemes could recover at first post-Newtonian order. However, Havas regarded the radiation effects as being the particular target of the method and expected that the scheme, which had presumably not been pushed to a high enough order (one past the linear order in the 1962 paper) to recover the perihelion shift, might still correctly derive the back reaction on a system due to wave emission.

At first, Havas, Goldberg, and Havas's student Stanley Smith all arrived independently at a result showing a loss of energy by such a system. In fact, Havas had derived this result as early as 1957, when he concluded that "the gravitational and the electromagnetic radiation damping terms are of the same form, and thus it appears that gravitational radiation effects have as much reality as electromagnetic ones"(1957). But during 1958 Havas noticed an error in his calculations that, as with Hu's (very different) calculation years before, reversed the sign of the result. Havas recalled that "all three of us had been so sure that there must be damping that we had not paid enough attention and each with a different slip had indeed gotten it" (pers. comm.). Smith and Havas, after thoroughly checking this "disquieting result," discussed it at length in a 1965 paper.

They first of all noted that it was contradicted by Trautman's result showing that the use of retarded potentials ought to lead to an outgoing flow of radiation. Given the unphysicality of their result, it appeared that the discrepancy was due to a failure in the approximation, at least to the order pursued by Smith and Havas (again, first postlinear). Therefore,

> although the possibility should not be overlooked that an approach to the problem of gravitational radiation by considerations of energy flux at infinity is inherently inadequate, we would rather expect that an investigation of the higher orders of approximation would indeed yield an energy loss in the retarded case (or possibly show the absence of any energy change). (Smith and Havas 1965, pp. B505–B506).

Havas's views on the analogy with electromagnetic waves are the most interesting of any skeptic, since they underwent considerable evolution. To begin with, one might say it was the prospect of applying his experience with the electromagnetic case that prompted him to take up the problem of the gravitational radiation reaction. By the time of his 1957 paper, when he still thought his calculations indicated a result with the "correct" sign, he concluded with a ringing endorsement of the analogy. However, experience convinced him that the analogy was on shaky ground and that faith in it was liable to mislead in the case of the gravitational problem, as he made plain in his 1965 paper with Smith. In the interim he had become the most exacting and detailed critic of the linearized approximation and the analogy it inspired, noting that the linearized approximation does not actually permit motion of the type that would emit radiation, because of the restrictions on motion placed by the linearized gravitational field equations (see chapter 6 for an earlier mention). He argued that it might well prove true that the exact equations essentially did the same thing, forbidding freely falling systems from following a path which would actually generate gravitational waves.

Despite a great deal of other work at this time on fast-motion approximations of the problem of motion (not primarily directed at the radiation problem in most cases) (e.g., Kerr 1959; Bertotti and Plebanski 1960; Westpfahl 1985), the result for the binary radiation reaction problem found by Havas and collaborators presented a problem for further work on the radiation problem in this approach. The first postlinear order was already rather complicated, and it would appear there was little stomach for attempting the next order. Havas thought that the method suffered from a lack of exposure, complaining at the GR meetings of that time that Infeld, who championed the post-Newtonian EIH method, prevented papers being given on its rivals (Havas, interview). Plebanski, who worked both in the EIH and fast-motion schemes reports that Infeld showed some hostility to the latter approximation method (Plebanski, interview).

Havas himself would have liked to press ahead with the next order of approximation but lacked the time and the manpower (in the form of students and collaborators) for such an undertaking. In the early '60s he benefited from his association with Goldberg at the USAF and was able to get air force grants to attract to Lehigh visitors like Plebanski and the Israeli physicist Moshe Carmeli. Such researchers looked at the problem of a massive particle moving in an external field and showed how the field could be divided into the external field and the particle's own self-field, despite the problems that the nonlinearities of the field introduced compared with the electromagnetic case. With the termination of air force funding in the early '70s (and from Havas's point of view possibly Goldberg's departure from the ARL), funding for science in smaller institutions especially was strongly curtailed. The result was that, for the next few years, the field was left to new efforts in the slow-motion approximation. Havas, despite the disappointment he faced in his own efforts, still viewed with grave, perhaps increased, skepticism the post-Newtonian efforts, which he suspected were justifying their means by their ends:

> It is perhaps even more unfortunate that the inconsistent approxima-
> tion used [Einstein's original linear approximation] led to a result
> which conformed so completely to expectations based on physi-
> cal "intuition" borrowed from electrodynamics; this has led to a
> ready acceptance of Einstein's [quadrupole] formula for the sup-
> posed energy loss and a neglect of critical study of its derivation. It
> also has tempted many authors to *justify* other (classical or quantum
> mechanical) approximation methods by their ability to reproduce [the
> quadrupole formula], which clearly is not a valid criterion. (Smith
> and Havas 1965, p. B504)

Like Bonnor, he regarded the problem of radiation reaction in a freely falling system as still open.

In the years after the Chapel Hill meeting, general relativity as a field continued to progress and grow. The series of GR conferences were officially inaugurated, the first being held at Royaumont, near Paris in 1959 and the second in Warsaw in 1962. The perception among physicists is that a period of debate and concentration on a topic will serve to clarify matters and improve understanding of the physics involved. By 1962 it would be fair to say that debate on the question of whether gravitational waves existed had largely ceased, but there was still a great deal of discussion as to whether binary stars would radiate them. Indeed, the host of the conference, Infeld, was of the opinion that they did not. Bondi, who, as we shall see, had with his collaborators done much to improve the understanding of wave propagation far from the source (see especially Bondi, van der Burg, and Metzner 1962; and Sachs 1962), insisted at Warsaw that the problem of radiation from a binary system was still unsettled (Bondi 1964). There were those, like Feynman, who viewed this lack of "progress" and the caution of the relativists with impatience. If a field did not make sufficient progress on a problem in a period of years, then it could fairly be taken as a criticism of the field itself. For this reason some relativists, even from the early days, were reluctant even to acknowledge that such a debate persisted. As early as the Chapel Hill conference, Feynman had been, as he reported to Weisskopf, "surprised to find a whole day at the conference devoted to this question" of whether gravity waves could carry energy (Feynman to Weiskopf, Jan 4–Feb 11, 1961, Feynman Papers, California Institute of Technology, box 29, folder 14). He was caustic in his appraisal of the discussions at the Warsaw conference, writing to his wife, in a well-known letter written while waiting for Sunday dinner to be served at the Grand Hotel, Warsaw during the communist era:

> I am not getting anything out of the meeting. I am learning nothing. Because there are no experiments this field is not an active one, so few of the best men are doing work in it. The result is that there are hosts of dopes here (126) [the number of participants in the conference, according to the published proceedings, was 114] and it is not good for my blood pressure: such inane things are said and seriously discussed that I get into arguments outside the formal sessions (say, at lunch) whenever anyone asks me a question or starts to tell me about his "work." The "work" is always: (1) completely un-understandable, (2) vague and indefinite, (3) something correct that is obvious and self-evident, but worked out by a long and difficult analysis, and presented as an important discovery, or (4) a claim based on the stupidity of the author that some obvious and correct fact, accepted and checked for years, is, in fact, false (these are the worst, no argument will convince the idiot), (5) an attempt to do something probably impossible, but certainly of no utility, which, it is finally revealed at the end, fails (dessert arrives and is eaten), or (6) just plain wrong. There is a great deal of "activity in the field" these days, but this "activity" is mainly in showing that the previous "activity" of somebody else resulted in an error or in nothing useful or in something promising. It is like a lot of worms trying to get out of a bottle by crawling all over each other. It is not that the subject is hard; it is that the good men are occupied elsewhere. Remind me not to come to any more gravity conferences! (Feynman and Leighton 1989, pp. 91–92)

Bondi's lecture, however, inspired the famous astrophysicist Subramanian Chandrasekhar to take up the problem (Chandrasekhar, interview). Born in India in 1910, Chandrasekhar is best known for his discovery, while on board ship to England from India in 1930, of an upper limit to the mass of white dwarf stars, above which such a star cannot avoid collapsing, crushing its own atoms. This result was extremely controversial in the

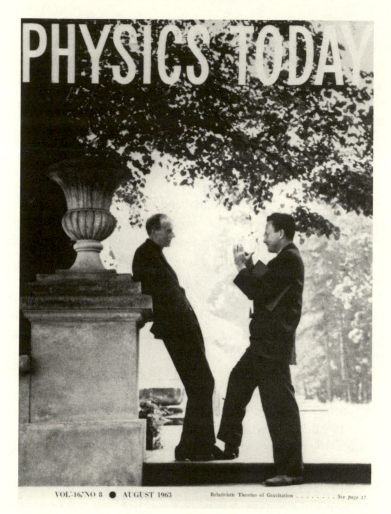

PHYSICS TODAY

VOL. 16, NO. 8 ● AUGUST 1963   Relativistic Theories of Gravitation . . . . *See page 17*

Figure 8.1. This photo of Richard Feynman (*right*) speaking with
Paul Dirac (*left*) at the Warsaw conference on General Relativity and
Gravitation (GR2) in 1962 made the cover of *Physics Today*, a sign of
the increasing importance of the field. Presumably Dirac was not one
of the 126 "idiots" Feynman encountered at the conference. (Courtesy
AIP Emilio Segré Visual Archives, *Physics Today* Collection)

thirties, and the hostility that Eddington showed towards it in public forced
Chandrasekhar to leave England for America, where he worked for the rest
of his life at the University of Chicago (see Wali 1992 and Miller 2005).
In 1983 he received the Nobel Prize in physics, principally because of his

celebrated work on white dwarfs. In 1962, inspired by a desire to work on general relativity just as the subject was about to take off, he gained permission to attend the Warsaw conference, the result of which was that he took up the problem of radiation damping of binary systems.

Throughout the 1960s, Chandrasekhar developed his own slow-motion formalism, dealing with extended fluid bodies (as opposed to point masses) at one post-Newtonian order after another (Chandrasekhar 1965). By the end of the decade he had advanced far enough in the expansion (to post-$2\frac{1}{2}$-Newtonian order) to describe reaction effects. His conclusion agreed with the quadrupole formula result (Chandrasekhar and Esposito 1970).

At about this time William Burke, a student of Kip Thorne's at Caltech, introduced improvements to the slow-motion approach that removed much of the arbitrariness in imposing the boundary conditions. Burke selected the problem of radiation damping in binaries for himself, since Thorne had been convinced by Peres's work that the problem was solved in the slow-motion case. Introduced to general relativity by Frank Estabrook of the Jet Propulsion Laboratory (later to become a pioneer of the use of Doppler tracking of deep-space probes such as the Voyager spacecraft to attempt to detect gravitational waves), Burke was greatly influenced by an applied mathematician at Caltech, Paco Lagerstrom (Thorne and Estabrook, pers. comms.). Lagerstrom and his group had developed the technique of matched asymptotic expansions (abbreviated MAX), which had finally set on a mathememantically secure foundation the brilliant insights of Ludwig Prandtl that inaugurated the modern field of fluid dynamics in the early twentieth century. Werner Heisenberg had said of Prandtl that he had "the ability to see the solution of equations without going through the calculations" (Narasimha 2004). Lagerstrom had provided a way to actually do the calculations. Burke saw that this revolutionary technique was the solution to the problem that plagued the radiation reaction problem in relativity. (Note that general relativity shares with fluid dynamics the challenge of nonlinearity in the equations.)

Matched asymptotic expansions is a method that permits one to compare two quite separate approximation schemes having different areas of validity by matching them term by term in an intermediate zone where they are both valid. Burke realized that doing so could allow one to compare a post-Newtonian expansion, valid for the source of the gravitational waves, with

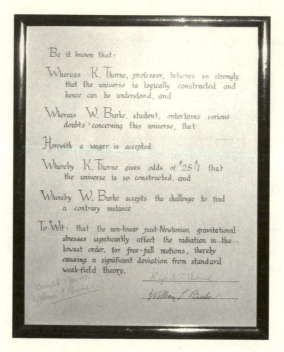

Be it known that:

Whereas K. Thorne, professor, believes so strongly
    that the universe is logically constructed and
    hence can be understood, and

Whereas W. Burke, student, entertains serious
    doubts concerning this universe, that

Herewith a wager is accepted

Whereby K. Thorne gives odds of $25:1 that
    the universe is so constructed, and

Whereby W. Burke accepts the challenge to find
    a contrary instance

To Wit: that the non-linear post-Newtonian gravitational
    stresses significantly affect the radiation in the
    lowest order, for free-fall motions, thereby
    causing a significant deviation from standard
    weak-field theory.

Figure 8.2. This framed copy of the bet made between Bill Burke and his advisor Kip Thorne hangs on the wall of the Bridge Building at Caltech. Like many skeptics before him, Burke hoped for a novel result hidden in the nonlinear complexities of general relativity. (Courtesy Kip Thorne and Michele Vallisneri. Photo by Michele Vallisneri)

a linearized approximation, valid towards infinity, where the waves actually existed, by matching them in an intermediate zone far, but not too far, from the source. In this way boundary conditions, such as outgoing-wave-only, that could be stated unambiguously in the linearized approximation could be imposed in a consistent way on the near-zone solution that actually described the source, thus addressing the arbitrariness that had bedeviled the slow-motion approach up to this time (Burke 1969). With this novel approach, Burke and Thorne also derived the quadrupole formula for emission from binary systems (Burke and Thorne 1970). Burke also constructed a radiation-reaction potential that could describe the damping force exerted on the orbiting bodies. Essentially it showed one how to construct the terms that would otherwise be missing from an expansion constructed using the half-advanced-plus-half-retarded potential.

Despite Thorne's complacency with that solution, Burke initially considered that his contributions did not settle the issue of whether bound systems would experience damping. He noted in early versions of his work that his approach was not guaranteed to work outside of linearizable systems and therefore could not settle the issue for freely gravitating systems. In yet another illustration that students do not always slavishly follow their mentors, one of the leading nonskeptics had a somewhat skeptical protegé. There is still on display at Caltech the record of a wager between Burke and Thorne on whether nonlinear effects would "significantly affect the radiation in the lowest order" from sources in free-fall motion. Thorne gave odds of 25 to 1 for this bet, which Burke conceded in 1970 (see figure 8.2).

If one looks at the bets Kip Thorne has made over the years (as I mentioned previously there are several posted outside his office at Caltech), one notices that he hardly ever loses bets about physics but rarely wins bets about how long it will take to accomplish a given objective. For instance, he has lost more than one bet about the date by which gravitational waves would have been detected. The bet with Burke nicely illustrates the phenomenon, in that Kip won the bet by 1970 and yet would, I suspect, have been astonished to learn that the following decade would be marked by intense controversy over the validity of the quadrupole formula for sources in free-fall motion. His own student might have conceded, but many others were far from ready to do so.

9

Portrait of the Skeptics

Like all stories, this book gives an account from one perspective. But there are other perspectives. Some of the people who worked on this problem will hardly recognize the story I am telling or will think it unbalanced. The reason is that the physics community is actually many different communities which overlap each other. Until very recently the relativity community was a somewhat isolated and marginal one within the broader physics community. Most physicists would have found the concerns expressed by the skeptics eccentric and even incomprehensible, if they ever heard them. Within the relativity community there were those who were impatient with the skeptics' concerns (like Feynman), those who were faintly embarrassed by them (what would the other physicists think if they could see us arguing over this?), and those who simply ignored them. Of course there were also many who were sympathetic to them. Ignoring each other is something that physicists do all the time. Generally, to read someone else's paper is the exception, not the rule. There are far too many papers, even in a small field, to read all of them, and most physicists don't much like to read. They prefer to do their own research. Bondi, to give one example among many, told me he was notorious for not reading papers; he preferred to talk to people about their work. Infeld tells us that Einstein would never look something up in the literature, preferring always to work it out for himself:

> Einstein's habit of working everything independently is carried to the extreme. Once when we had to perform a calculation which was quoted in many books I suggested: "Let us look it up. It will save time." But he proceeded with his calculations, saying: "It will

be quicker this way. I have forgotten how to use books."(Infeld 1941, p. 277)

While every physicist has some colleagues whose papers they read, they generally don't read too widely. The physicist who reads every new paper in even one journal, such as *Physical Review*, is more admired than emulated. Many prefer to talk to an expert if they are unable to work the problem their own way, and even then they talk to the people whose opinion they really care to know. It was certainly quite possible for some relativists, even if they were interested in the radiation problem, to barely be aware of the sort of things the skeptics were saying.

There is a story from my own days as a student that illustrates this point. It was during graduate school that I did most of the research presented in this book, and just about the time I was to receive my degree, I gave a talk on my work at a conference attended by a great many distinguished physicists, including relativists. The conference was held to commemorate one of the most celebrated physicists of the last century, Andrei Sakharov. I was speaking at the conference only as the fortuitous result of a research trip to interview one of these grand old men. On that trip I had the pleasure of visiting John Wheeler in Princeton, one of the most distinguished men in the field of relativity. Wheeler was the thesis advisor to my thesis advisor, Kip Thorne, so this visit is especially memorable for me. As it turned out, the great man was not terribly interested in talking about the past and instead of me taking out my tape recorder and asking him questions, he took me for a stroll around and around the corridors near his office while quizzing me about the latest work in Kip's group at Caltech. Earlier, when I first arrived at his office, I had found him dictating a letter to the organizers of this Sakharov conference (actually the Second International A.D. Sakharov Conference on Physics), who had apparently asked him to nominate some speakers whom they should invite. In what I later learned was a typical Wheeler gesture, my appearance immediately moved him to add my name to the list, to talk about the history of gravitational waves. Wheeler was well known for sending his students to talk about their work at major conferences, rubbing elbows with the big names of the field. He believed that in this way young people would gain valuable experience and (no doubt) that the older people would learn something at the same time. (In his autobiography, Bondi says that when he was president of the

Royal Astronomical Society he would tell younger and less experienced speakers, who he felt were generally apt to overestimate their audience's ability to assimilate highly technical material, "You are addressing some very distinguished astronomers. Please speak to them as you would to twelve-year-old children"[Bondi 1990, p. 133]) Wheeler was a prolific mentor to young physicists, and after he began to do research in relativity, his students quickly formed one of the leading schools in the field, despite his relatively late entry into it.

Sure enough, the conference organizers could not ignore the advice of John Wheeler, and though they had surely never heard of me, I received a letter inviting me to speak at the conference at one of the smaller sessions. My own advisor, Kip, who models himself on his mentor Wheeler in his dealings with his students, gathered the required money to send me all the way to Moscow to attend. So I found myself giving my talk to an audience brimming with the very people who made the history I was trying to describe. I was introduced by one the leading figures in relativity theory, the chair of the session, Robert Wald. I began by saying I was going to discuss why so many relativists were skeptical of gravitational waves and their sources in the 1950s and was immediately interrupted by Bryce DeWitt, sitting in the front row:

> You weren't there in the 50s. [Undeniably true!] There was no one saying then that gravitational waves didn't exist.

I can't guarantee these were his exact words, but they were fairly close. The occasion has somehow stuck in my mind. This put me in a little bit of a quandary. I was about to marshal a host of supporting quotations from the typescript proceedings of the very conference (at Chapel Hill) that DeWitt had helped his wife Cécile organize in 1957. My transparency viewgraphs were littered with the citation to the proceedings, "DeWitt 1957." Furthermore DeWitt and DeWitt-Morette are among the biggest names in the history of radiation theory in relativity. Admittedly their own work focused on electromagnetic rather than gravitational waves, so I had never tried to interview DeWitt, but obviously he ought to know better than I. How was I going to tell the man himself that my reading of the proceedings of his wife's conference contradicted his own recollection?

Fortunately I didn't have to. One of the men I had interviewed, one of Wheeler's most famous students, Charles Misner, spoke up from the back and said "Don't you remember Bryce that Bondi was saying so and so . . . ?" A distinguished Russian physicist, Leonid Grishchuk, also spoke up in my defense, recalling Infeld's skepticism, which was well known in Russia. DeWitt then allowed that, now that Misner mentioned it, he did recall that there was some such talk, and they began to debate the matter between themselves. Fortunately, at this point, the chairman intervened on my behalf and suggested that I should actually be allowed to give my talk before, rather than after, the group discussion. I am glad to say that, by the time I was finished, no one found too much to quarrel with in my account, but I tell this little anecdote to illustrate one way in which this book tells a story that is not entirely recognizable to some of the people who were there. In focusing on the question of whether gravitational waves exist at all, or are ever emitted by binary stars, I am naturally drawn to a perspective that emphasizes those who were skeptical on this point and that tends to exclude those who never regarded it as a open question.

Of course, within the historical memory of the field itself an opposite tendency is at work. Whereas my account tends to overemphasize the debate about whether gravitational waves existed simply by focusing on it, the folk-memory of modern relativists tends to underemphasize it. While the temporary or aberrant skepticism of such influential figures as Einstein and Eddington is preserved in the folk memory in occasional anecdotes or quotations, the suggestion that there was ever any real *debate* on such a subject is frequently rejected or resisted. Individuals may have made errors at times or held erroneous views, but to suggest that there was ever much public discussion or that there was really a problem in the general sense would be to go too far. There is a preference not to remember or not to overstress the significance of something which may be seen as vaguely disreputable to the field. It is a characteristic aspect of physics that to pose a problem or a question may, in itself, be taken as a sign of bad character. It is typical of an established theory framework that certain questions become nonsensical and certain problems become redundant when a new paradigm becomes established, even though they were once perfectly reasonable issues. Referring to Newton's success in answering dynamical questions about the solar system (what are the forces that keep the planets in motion?) while ignoring evolutionary issues (how did the planets get to be in the orbits they currently occupy?), which had previously seemed all part of the same problem, Bondi saw such change as "a vital feature in the whole pursuit of science"(1970, p. 266)

In looking at the skeptics and their concerns, I want to try develop some insight into how physics is done by looking at cases where every step in a problem is open for debate. Given this, it is time that I tried to present a portrait of some of the skeptics themselves, which confronts me with the issue of how to define a skeptic. Broadly speaking, several relativists are remembered as having been skeptical of the existence of gravitational waves. Of course, their views were far from monolithic, especially because they were always a minority. Some people did in fact suggest that gravitational waves do not exist, for instance, Einstein and Rosen, both rather briefly. More common was some variant of the view that radiation damping of orbits does not exist, in the sense that gravitational waves do not draw energy from gravitational systems such as binary stars and/or do not transport energy. Into this category we may conveniently sweep Infeld, Bondi, and Rosen, among others.

An alternative form of skepticism concerns the detail of the back-reaction problem, specifically the quadrupole formula which describes (in the orthodox view) the rate of emission of gravitational wave energy from a freely falling system. Notable skeptics of this sort, including Havas, Ehlers, Rosenblum, and others, can be categorized under a different use of the word. They are skeptics in the sense of being agnostic about a subject. Most of the previous group (but not all) had an alternative view of the nature of gravitational waves in mind. While most of the critics of the quadrupole formula did not offer any particular formula to replace it, they more typically had a precise view of how it ought to be derived. The two forms of skepticism could easily overlap. To say that the quadrupole formula might be incorrect implied the possibility (frequently stated by Havas and stressed also by Eddington and Bondi) that there was no quadrupole emission at all, so that back reaction in gravity could be much weaker than was thought, or even nonexistent. As I have already argued, I look upon all of these types of skepticism as linked by the common thread of skepticism towards the electromagnetic analogy itself, an analogy which naturally tended to encourage the expectation that gravitational waves do exist, can transport energy, and are emitted by accelerating masses such as binary stars in the lowest order of emission permitted by the tensorial character of the gravitational field (i.e., the quadrupole order).

A curious duality may be observed here concerning attitudes to the idea of skepticism. On the one hand, the skeptic may be viewed as the ideal type of scientist, one who has no time for received opinion unsupported by factual evidence or experimental data. But as time went on in the study of gravitational waves, increasing signs of impatience could be observed, in some quarters, at the persistence of the skeptics. We shall later encounter Feynman's views on the importance of optimism and the necessity for a progressive research program that sets aside cavils or doubts at the outset and presses ahead until either all problems are overcome or the doubts are vindicated by irreconcilable internal contradictions. One can perhaps see here an epistemological struggle over the proper balance between skepticism and belief in theoretical science. Gravitational radiation, a field with no experimental input whatever for several decades of its development, was certainly an ideal arena for such a discussion.

In the context of a small field, and an even smaller subject area, any minority position must depend for its survival on a relatively small cast of characters. The fortunes of these persons and their success at passing on their views must considerably affect the chances of a successful challenge to the orthodoxy. How influential were the skeptics? Were they outsiders in their field? In a professional caste where it is easy to find oneself labeled a crank, were they able to command tolerance for their opinions? It is important to keep in mind the fluidity not only of the orthodox position but also of the views of individuals. There can be no hard and fast definition of a skeptic in this field or any other, but we will make do by following, as much as possible, the self-definition of the individuals involved. Rosen several times wrote or co-wrote papers whose titles questioned the existence of gravitational waves. Bondi also went into print, in a letter to *Nature* in 1957, describing himself as a (temporarily) convinced skeptic. This chapter is not meant even as a scientific biography of the three men whose work on gravitational waves I will discuss. All of them worked in many other areas in addition to this. Nevertheless Bondi, the best known of the three, has said he was particularly proud of his work on gravitational waves (Bondi 1990, p. 79). It was a subject that occupied them over the years, even when it must have seemed a strange one in the eyes of many other physicists. After all, when would gravity waves ever be detected? Probably the most prominent and outspoken skeptic was Einstein's collaborator, Infeld, so we will begin with him.

Leopold Infeld was born in Krakow in 1898, a part of Poland then ruled by the Austrian empire (Pyenson 1978). He received his doctorate from the Jagiellonian University (also Nicholas Copernicus's alma mater) in his home town in 1921, but in the new Polish republic there was little academic opportunity in physics, especially for a very young, Jewish physicist. After spending some time as a rural schoolmaster, he finally achieved the position of docent at the University of Lwów. In 1933 he obtained a Rockefeller fellowship, which permitted him to travel abroad, to work in Cambridge with Max Born and at Leipzig with Bartel van der Waerden. Upon his return to Poland, there was still no prospect of academic advancement, despite the publication abroad of some well-received papers, so he left once again, to take up another fellowship working with Einstein in Princeton, New Jersey. Here, no doubt, he was able to take advantage of his connection with Einstein's close friend, Born.

His arrival in late 1936 coincided, as we have seen, with the later stages of Einstein's belief that gravitational waves could be shown not to exist. Despite the disappointment of the subsequent failure of this position, the collaboration that followed was the making of Infeld's career as a physicist. The EIH paper became the canonical first post-Newtonian solution of the two-body problem of motion in gravitation. Its method permitted much more general calculations of relativistic corrections to traditional problems in orbital mechanics. In fact, the paper was immediately followed in *Annals of Mathematics* by a paper of H. P. Robertson's recalculating, on the basis of the EIH results, the famous perihelion shift of Mercury due to general relativity.

Although the EIH paper made a name for Infeld in physics circles, his close friendship with Robertson and the public side of his collaboration with Einstein may have been even more responsible for his subsequent professional success. After a year at Princeton, Infeld's fellowship ended, before the completion of the EIH research. Anxious to continue his collaboration with Einstein, Infeld suggested to his mentor the project of writing a popular book together, the proceeds of which would pay Infeld's wages for another year. Einstein, who clearly valued Infeld as a collaborator, readily agreed. Their book, *The Evolution of Physics*, was a best seller and received widespread publicity. From then on Infeld had a public profile as Einstein's collaborator, a status not achieved by any of the other physicists who worked closely with the most famous scientist of the age. Despite this, Infeld seems to have more or less given up the thought of securing a professorship at this stage. He had turned forty and had thought of trying to make a living writing popular science books (Infeld 1941).

Figure 9.1. Leopold Infeld (at right), at a postwar physics conference, seen with his former mentor, Max Born. Infeld became one of the leading figures in General Relativity in the 1950s and 1960s. (Courtesy AIP Emilio Segré Visual Archives, Gift of Jost Lemmerich and the Born Collection)

However, Robertson, with whom he had become very friendly at Princeton, made efforts on his behalf and persuaded John Synge, the relativist and mathematical physicist who was at the University of Toronto, to consider Infeld for a position there. Infeld went to Toronto as a temporary lecturer for one year and did secure a permanent appointment at the end of it. Synge, originally from Dublin, was head of the applied mathematics department at Toronto, which had been created for him. In Canada, at this time, theoretical physics was not a separate field. When, during the war, Synge left for Ohio State University, his department and Infeld were merged once more with the mathematics department. Infeld worked hard to found a center for theoretical physics in Toronto, but although he produced many students, he could not persuade the university to create new positions. Since there was no other center for the study of relativity in Canada, most of his students moved into other fields of physics after graduation. Nevertheless, Infeld, in spite of the university's lack of enthusiasm for the project, did a great deal to promote the field of theoretical physics in Canada (Wallace 1993).

When Infeld first went to Toronto, he worked with a student, Phillip Wallace, on generalizing the EIH method to electrodynamics. In this paper he had some interesting things to say about back reaction, in view of his later work, but it was not until after the Second World War that he really turned his attention to this problem in the gravitational case. In the meantime he finally persuaded his department to retain one of his students, Alfred Schild (a war refugee interned in Canada by the British as an "enemy alien" along with Hermann Bondi), after graduation but could not manage to get them to pay enough to keep him for long. Schild left for Pittsburgh and the Carnegie Institute of Technology (where Synge had also spent time after the war), taking another young student, Felix Pirani, with him. With Schild, Infeld had worked on the motion of a massless "test particle" moving in an external gravitational field (Infeld and Schild 1949).

In the late 1940s, Infeld returned to the EIH formalism with another student, Adrian Scheidegger, and addressed the problem of radiation reaction. They concluded that the problem of motion for gravitational binaries allows for no dissipation of the system's energy by radiation. They published a paper to this effect, and Scheidegger addressed both the American Physical Society and a conference in Vancouver on the subject. The talks did provoke some reaction. Scheidegger spoke laconically of a "considerable flow of discussion" after his talk at the APS (1951, p. 883), and Peter Bergmann (who was at the APS meeting), had one of his students, Joshua Goldberg, respond to the Infeld-Scheidegger assertions with a paper of his own.

As fate would have it, however, neither Infeld nor Scheidegger was able to continue the debate for long at this time. In 1950 Infeld had the intention of spending a sabbatical year in his native Poland in an effort to help rebuild the physics community in that war-shattered country. A small, Catholic, Canadian paper, the *Ensign*, chose this occasion to launch an attack on him, based on the ludicrous assertion that Infeld intended to give away atomic secrets to the Soviet Union or its ally the People's Republic of Poland. Infeld's close association with Einstein was farcically presented as evidence of his familiarity with nuclear-weapons secrets. In fact, neither Einstein or Infeld had been in any way involved with the nuclear-weapons research of either Canada or the United States. The mainstream press took up this slanderous attack, sometimes under the guise of reporting Infeld's denials, and the campaign reached its peak with a personal attack launched under the protection of parliamentary privilege by the leader of the opposition Progressive Conservative Party, George Drew.

Drew went so far as to demand that Infeld be prevented from leaving the country. The University of Toronto came under pressure to deny Infeld leave of absence. This had the effect of forcing Infeld to choose between Canada and Poland. In the light of the public and personal attacks against him and under surveillance and possible harassment by the Royal Canadian Mounted Police, he chose Poland. Rather suddenly, therefore, his career in Canada came to an end, and he was forced to begin again in Warsaw.

The motivations for the attacks made against him may have been various. He was an outspoken critic of American and British nuclear policy of the day, having stumped the country speaking to impress upon the Canadian public the futility of attempting to keep the "secret" of the bomb from the Soviet Union's physicists. He himself was a socialist and his American wife was also very left-wing and had involved herself to some extent in progressive Canadian politics, opposing Drew during his tenure as Ontario's provincial premier. Infeld had publicly joined in the defense of those accused in the Gouzenko affair, in which a defector from the Soviet embassy in Ottawa had denounced a number of prominent Canadians, including several scientists, some of whom were tried for espionage (see Reuben 1955 for more on this affair). The likeliest explanation is that he was an attractive target for the politically ambitious Drew and the hysterically anticommunist Catholics of the *Ensign*. The charges were absurd, since the Soviet Union had already exploded an atomic bomb. That fact was held against Infeld, with amazing chutzpah, by Drew, who regarded it as a suspicious fulfillment of Infeld's own prophecy that the Soviets would get the bomb in spite of Western efforts to keep it secret. The illogic of preventing someone from traveling abroad to give away secrets that he was insinuated to have already betrayed was not noted by the press until after Infeld's, largely involuntary, "defection." Likewise, the Canadian defense research establishment only then went on record to say that he held no military or scientific secrets. In fact, Infeld's war work was limited to some efforts with Synge in ballistics and some radar work. His former student Wallace did work on the British-Canadian bomb project, which was the closest connection Infeld had to the bomb. Wallace has insisted that the charges against Infeld were totally without foundation (Wallace 1993; for more general insights into cold-war suspicions of well-known physicists, see Kaiser 2005).

After Infeld's departure, Scheidegger continued to advocate the view that gravitational back-reaction did not exist in general relativity, exchanging papers with Goldberg in the *Physical Review*. But, like so many of Infeld's students in Canada, he found it difficult to secure an appointment in relativity or in theoretical physics. After a few years he found employment with an oil company in Canada and subsequently took up a successful academic career in geophysics. With that, the first round of debate on the topic of radiation reaction gradually petered out.

Infeld, despite the misfortune that had forced him to uproot his Canadian family and return to his homeland (in a shameful act, his two Canadian-born children were later stripped of their Canadian citizenship, as was Infeld himself, by the extrajudicial maneuver of "orders in council"), found his professional goal of establishing a school of theoretical physics easier to achieve in Poland than in Canada. The Polish government was eager to rebuild the country's scientific and educational infrastructure, which had been ruthlessly destroyed by the occupying Nazis. With personal fame as Einstein's collaborator and notoriety as a refugee from political persecution in the West, Infeld, became one of the leaders of this rebuilding effort. He made a considerable success of the opportunity, establishing a thriving school of relativity and producing, as before, many excellent students.

His political troubles did not entirely cease with his move to a communist country. His arrival coincided with the last years of Stalinism, and his association with Einstein was not an unmixed blessing in a political environment in which Einstein was still considered an idealist opposed to the true practice of Marxist science. The fact that Infeld publicly opposed the arguments of Vladimir Fock, who wished to reform general relativity by removing general covariance, did not help matters either. But Infeld did not, like some others, come under significant pressure to recant, and after Stalin's death in 1953, the political environment in Poland softened rather quickly.

In Warsaw, Infeld at last had a thriving group doing excellent work in general relativity, one of only a handful of relativity groups in the world at that time. Furthermore, from 1955 on, gravitational waves and the problem of back reaction became an active topic in the field. It is interesting that Infeld's collaborators in Poland did not share his views on this subject. Andrzej Trautman and Jerzy Plebanski did important work on the problem of motion and gravitational waves, and both were decidedly nonskeptics. Yet throughout this period (the late '50s and early '60s), Infeld persisted in his own viewpoint, as shown in his 1960 textbook *Motion and Relativity*, written with Plebanski. Infeld displayed a certain high-handedness in dealing with his co-author's difference of opinion. The chapter of *Motion and Relativity* dealing with radiation reaction was added to the book without Plebanski's knowledge, after the latter's departure for America on a fellowship. The chapter is an excellent account of Infeld's position at the time but entirely fails to represent Plebanski's view, which was diametrically opposed. Similarly, in one preprint, Infeld alluded to the contrary views of his students but asserted that they had come to accept his arguments as correct. This did not stretch the truth so much as contradict it entirely.

Nevertheless, despite these public vanities, Infeld was personally fair with his students. Plebanski, whose family suffered severe repression under the communist government of Poland (his uncle died of ill-treatment suffered while under arrest for his political activities), wished to leave Poland for good, and Infeld helped him in securing permission to move "temporarily permanently" to Mexico City. Trautman believes that personal difficulties between Infeld and Plebanski in the wake of the *Motion and Relativity* affair played a role in Plebanski's desire to leave Poland (pers. comm.). Trautman, who was a very prominent contributor to the new advances in the picture of gravitational waves, including an important calculation of the radiation reaction effect that supported Goldberg's position in the old debate with Infeld and Scheidegger (Trautman 1958b), remained an important and favored member of Infeld's group. Shortly before his death, however, Infeld seems to have been finally won over by his students' arguments, rather than the other way around, and even published a paper with Róża Michalska-Trautman (wife of Andrzej Trautman) that accepted the existence of the phenomenon of radiation reaction (Infeld and Michalska-Trautman 1969). In this case, therefore, the advisor did not perhaps influence the students so much as they influenced him, though there have also been suspicions that Infeld's opinions were not accurately reflected in this late work, an ironic reversal (if true) of the earlier story with Plebanski. In spite of Infeld's great personal success as an advisor and the founder of a school, this skepticism was not at all transmitted to his students who remained active in the field. Nevertheless, his interest in this subject was passed on and led to much significant work that developed the study of gravitational waves in the 1950 and '60s.

Nathan Rosen, Einstein's other collaborator directly involved with the attempt to disprove the existence of gravitational waves, also remained a prominent skeptic of gravitational waves throughout much of his career. Born in 1909 in Brooklyn, New York, Rosen was, like Infeld and many other scientists of the time, a socialist. He was so strong in his convictions as to wish to live and work in the Soviet Union at a time, the late 1930s, when that country's suspicion of foreigners was at its height, and it required Einstein's intervention to secure for him a position at the Kiev State University. Whatever his opinion of "actually existing socialism" during the era of the great purges, he returned to the United States after only two or three years. His letters to Einstein during this period are full of enthusiastic praise of the Soviet system, right up to the moment, on July 31, 1938, when he abruptly announced his intention of returning to the United States permanently. Reading these letters half a century later, it is difficult not to wonder whether the praise is a little forced. Certainly foreigners in the Soviet Union during the purges were in great physical danger, since many of the accusations made at the famous show trials of this period revolved around allegations of sabotage and "wrecking" by foreigners. Given the hysteria that was undoubtedly in the air and the very real prospect of being arrested out of the blue, Rosen and others, like the Polish physicist Mathisson, found it necessary to leave at this time. In the

Figure 9.2. Nathan Rosen and Joseph Weber, key figures in the controversies surrounding gravitational waves in both theory and experimental communities. Despite his iconoclasm, which included developing an alternative theory of gravity in rivalry with GR, Rosen remained a highly regarded figure in his field for his ready willingness to admit when his ideas had not proved correct. (Courtesy AIP Emilio Segré Visual Archives)

1950s, again with Einstein's endorsement, he emigrated to Israel and built up the physics department at the Technion Institute in Haifa, where he worked until his death in 1995.

Rosen might be fairly viewed as a professional skeptic in the best sense and has played a prominent gadfly role in the history of twentieth-century theoretical physics. His name has been immortalized in that role because of the famous paper he wrote with Einstein and Boris Podolsky, which formulated the EPR paradox in opposition to the Copenhagen interpretation of quantum mechanics. Although he agreed with Einstein that there was a difficulty with their original paper on gravitational waves, he felt strongly enough about the argument to publish in a Soviet journal a revised version that restricted itself to disproving the existence of plane gravitational waves (Rosen 1937), an argument that was rebutted after the war by Bondi, Pirani, and Robinson (1959).

In 1955, at the Bern jubilee conference, he turned to casting doubt on the reality of the cylindrical wave solution from the published version of the Einstein-Rosen paper. He suggested that these might not carry any energy (based on an analysis of the energy pseudo-tensor in cylindrical coordinates) but subsequently retracted this view (Rosen 1958). Afterwards, he was prevented, possibly by institutional commitments at the Technion, from personally pursuing further work on gravitational waves for many years (Peres, pers. comm.). Indeed, his final correction of the Bern paper did not appear until 1993 (Rosen and Virbhadra 1993), an unusually long publication delay by anyone's standards!

However, in 1979, perhaps inspired by the resurgence of interest in the problem at that time, he published a paper that returned to the problem of the arrow of time in gravitational radiation theory. The title of this paper, "Does Gravitational Radiation Exist?" (Rosen 1979), echoed that of his 1936 submission to the *Physical Review* with Einstein (Einstein and Rosen 1937). In the new paper he adapted the Wheeler-Feynman absorber theory to gravitation and concluded that as the gravitational force interacted much less strongly with matter than the electromagnetic field, a source system would not undergo radiation reaction for lack of a sufficiently strong absorber field. In the Wheeler-Feynman theory it is the field of the absorbers, back-reacting on the source, that breaks the time symmetry of the source field (Wheeler and Feynman 1945, 1949). Regardless, Rosen's arguments do not appear to have completely convinced even himself, since towards the end of the paper he retreated to a more Tetrode-like position, conceding that an absorber (such as a gravity wave detector) could presumably act so as to draw energy from the source at a distance. In any case, his paper did not excite much debate on the subject.

Nevertheless, in the 1950s he had been indirectly responsible for one very important contribution to the back-reaction problem when he encouraged his student Asher Peres to pursue the problem, in an effort to decide between the rival views of Fock and Infeld on the existence of radiation damping in freely gravitating systems. Peres developed an approximation scheme that correctly reproduced the EIH and Fock results at first post-Newtonian order but then ran into difficulties calculating the radiation effects on the motion. His thesis results found, as had Hu, an energy gain by the orbiting system, but he subsequently located the source of his error in a failure to correctly apply no-ingoing-wave boundary conditions to the near zone of the system. By careful attention to matching the far-zone conditions to the near-zone solutions, he overcame this difficulty, much to his and Rosen's relief, and rederived the quadrupole formula.

Surprisingly, his success had relatively little impact, even though his paper was viewed very favorably by some experts, such as Kip Thorne. Although he had identified a key reason why previous calculations had produced such wildly varying results, he could not overcome the sense of dissatisfaction that the slow-motion approach had generated amongst some experts (e.g., Bonnor and Havas), and it was not until Burke introduced his general technique for matching far-zone and near-zone solutions that an unambiguous method of imposing the boundary conditions in the slow-motion expansions encouraged greater confidence in that type of approximation scheme. Nevertheless, Peres's 1960 paper, at least in hindsight, can be seen as a turning point for the back-reaction calculations. Previously it was difficult to find two results that agreed with each other. Subsequently, the great majority of slow-motion calculations agreed with the quadrupole formula in their predictions, at least for binary systems in circular orbits.

If Rosen himself had widespread influence on the field of gravitational radiation, it was perhaps by example. None of his published ideas were ever taken up except in rebuttal, but his gadfly role (e.g., he was the originator of a prominent rival theory of gravitation to general relativity) has attracted admirers, such as the most prominent current skeptic on gravitational waves, Fred Cooperstock. Rosen always sought to be provocative throughout his career and, unlike many scientists, was not afraid to stick his neck out with unconventional ideas and views in an effort to challenge received opinion and unwarranted assumptions. He continued actively researching literally up until his death (I myself refereed a paper of his only shortly before he died) and remained remarkably true to his own convictions in a profession in which fear of nonconformity occasionally dissuades people from publishing their best work.

Another fruitful skeptic was Hermann Bondi, one of the originators of the steady-state theory of cosmology. Though now with few remaining supporters, this was, in the fifties and sixties, a strong rival to the the big bang theory (see Kragh 1996 for an account). Bondi adopted Eddington as his mentor in the field of relativity and emulated his skepticism in regard to existing formulations of gravitational wave theory. Bondi, furthermore, took an independent stance towards what was worthwhile in general relativity. He disliked work of the problem of motion type, producing minor corrections to the Newtonian theory, and saw in gravitational waves an opportunity to study an entirely new phenomenon predicted by the theory that was unknown to Newtonian gravity and that might yield insight into the novel nonlinear aspects of the theory.

Bondi grew up in Vienna between the wars but chose not to pursue his education in physics in Austria during the period of the Christian Social Party dictatorship. Instead he went to England, where he secured placement as an undergraduate at Cambridge with the help of a recommendation from a relative who was a prominent mathematician, Abraham Frankel. At Cambridge he was an almost immediate success as a student, but the greatest advantage of his move abroad was shown shortly afterwards, in 1938, when Hitler invaded Austria. Acting on Bondi's advice, his family precipitately left Austria shortly beforehand and thus avoided the fate shared by many other Austrian Jews under Nazi rule. But exile did not immediately rescue Bondi from the class of suspicious persons by reason of his nationality. In England, once war with Germany broke out, he became an "enemy alien," and one of the first acts of the Churchill government in 1940 was to order his internment and that of many others like him. Nevertheless, in his autobiography, Bondi recalled his relief at Churchill's accession to power, owing to the latter's association with a policy of confrontation with Nazi Germany (Bondi 1990, p. 27).

An internment camp cannot be a pleasant place in which to find one-self, no matter how awful the alternative. Bondi was obliged, with other European exiles, to spend more than a year in detention in Canada, to which they were deported, despite the perils of predation by U-boats, which had sunk one unescorted vessel full of interned refugees just prior to Bondi's crossing. During his internment, Bondi encountered two other important theoretical physicists, incidents that illustrate the amazing intel-lectual quality of the European refugee population. One of the scientists was Thomas Gold, his longtime collaborator, and coauthor of the steady-state theory; another was Alfred Schild, who stayed in Canada after his release and became a student of Leopold Infeld's in Toronto. Schild would later cofound the immensely influential series of Texas symposia, which would witness both the early stages of the revival in general relativity research, and many years later, the announcement of the first experimental evidence for the existence of gravitational waves.

Despite this unfortunate interruption of his academic career, Bondi was determined to return to England (his family had immigrated to America before internment took effect), where he would resume his academic career at Cambridge and participate in war work with other rehabili-tated "enemies" such as Gold. Aided by the marvelous English facility for overlooking any unpleasantness one may have caused, he assimilated perfectly to English academic life and was eventually knighted for his later administrative work in the British defense ministry.

Bondi's interest in gravitational waves was sparked initially by the 1955 Bern conference commemorating the jubilee of special relativity. It was at this meeting that Rosen presented his paper suggesting that cylindrical gravitational waves could not carry energy. Following the Infeld and Schei-degger work of a few years earlier, the possibility that gravitational waves did not exist was in the air. Bondi himself recalled:

It [the Bern meeting] was particularly memorable for me because of the discussions we had . . . on gravitational waves. The mathematical and physical complexity of Einstein's theory of gravitation is so great that there was still confusion, and a variety of opinions, about whether the theory predicted the existence of gravitational waves or not. After one of these discussions, Marcus Fierz, professor at the ETH, the federal technical university, took me aside and said, "the problem of gravitational waves is ready for solution, and you are the person to solve it." This remark governed a sizeable slice of my scientific work for many years, and led to the 1962 paper on gravitational waves in a fifteen paper series. . . . The 1962 paper [presumably Bondi, Van den Burg, and Matzner] I regard as the best scientific work I have ever done, which is later in life than mathematicians usually peak. (Bondi 1990, p. 79)

Another factor that encouraged Bondi's interest in gravitational waves was the interest of his student and colleague Felix Pirani. Emigrating to Canada during the war from his native England, Pirani began his physics career as, very briefly, a graduate student of Infeld's at Toronto, but when Schild left Toronto to take up an appointment at the Carnegie Institute of Technology in Pittsburgh, Pirani went with him and completed his graduate studies there. Owing to Schild's friendship with Bondi from internment together, he then went to Cambridge, where he received a second doctorate working with Bondi on cosmology. Subsequently, after a year in Dublin with Synge (another Toronto and Pittsburgh connection) at the Dublin Institute for Advance Studies, he took up an appointment at King's College, London, where he formed part of a very active and influential group in relativity along with Bondi, who became professor of applied mathematics there in 1954.

Pirani first had his attention drawn to the confused state of gravitational wave theory when he was asked to review a paper on the subject for *Mathematical Reviews* (Pirani 1955). The paper, by the British cosmologist George McVittie (1955), who had been a doctoral student of Eddington's, attempted to show that plane gravitational waves could not exist, a view previously propounded, with different reasoning, by Rosen. Pirani felt that the result must be wrong and subsequently joined forces with Bondi and Ivor Robinson to write the canonical paper on plane gravitational waves. He benefited from his time in Dublin, where he was influenced by Synge in two important respects. The first concerned the equation of geodesic deviation, a coordinate invariant way of looking at physical interactions in general relativity, based on the curvature tensor and pioneered between the wars by Synge. Pirani was led to describe the interaction of a wave with a physical system by showing that the particles in the system would be moved about *relative to each other* by a passing wave. This description not only helped give a better picture of how a gravitational wave worked in practice, but also led to ways of sidestepping the vexed question of whether such waves carried energy or not, and so could or could not do physical work.

The second idea to which Pirani was introduced in Dublin was the classification of radiation fields by type. He was asked to proofread a book of Synge's that discussed an analogous ordering of electromagnetic fields and was inspired to do the same for gravitational radiation. He then came across a classification scheme based on types of the Riemann tensor for different fields, due to Petrov. This scheme, dividing gravitational fields initially into type I, II, and III, with the latter two describing radiation fields, was adopted and became widespread. The success of this scheme, after the failure of the earlier Weyl-Einstein attempt at classification, illustrates yet another irresistible impulse, to classify, which physicists (and other scientists) are subject to.

Whereas Pirani was skeptical of the skeptics, Bondi was influenced by Eddington, from whose book he learned relativity, to adopt his own brand of skepticism of gravitational waves. To begin with, he seems to have been doubtful whether relativity theory really admitted them. In his letter to *Nature* of 1957, which is remembered by many for its critical thought experiment "proving" the likely existence of gravitational waves, he described himself as having been a strong skeptic at Chapel Hill that same year. Joshua Goldberg, who helped facilitate that meeting and attended it, recalls that Bondi advocated the nonexistence of gravitational waves at that meeting (Goldberg 1993). Yet the contradictions between various remembered histories is illustrated by Pirani's remark, "I'm surprised that there was still some doubt on gravitational waves carrying energy at Chapel Hill" (interview)! Indeed, the famous thought experiment that was used repeatedly in 1957 to circumvent Rosen's pseudo-tensor problem was "enabled" by Pirani's groundbreaking work on the geodesic-deviation description of gravitational waves, which he reported at that conference.[1]

Collaborating extensively with other groups in Poland, Germany, and America and helped by USAF funding secured through Goldberg, the London group published a string of papers in the late '50s and early '60s that had a profound impact on the theory of gravitational waves. Few can have done as much as Bondi did to establish the current orthodoxy on gravitational waves and to encourage belief in their existence. Nevertheless, Bondi himself remained true to his skeptical roots. In 1962, at the Warsaw conference (GR3), organized by Infeld's group, he still regarded the question of whether binary systems were damped by radiation as open and felt that it was important to examine the case of two extended bodies with a real equation of state, however idealized. Chandrasekhar was inspired by this talk to take up the problem (interview).

Bondi thought that two orbiting bodies consisting of pressure-free dust would not radiate, since every particle contained in the two bodies would follow a geodesic throughout their motion. In an Aristotelian sense, these particles would behave "naturally" during their motion, and so could be regarded as nonaccelerating, in the geodesic sense, and therefore nonradiating. Bondi remained unsure whether damping should be present in this highly idealized system for many years (interview).

From the mid-sixties on, Bondi became increasingly involved in administrative work, a fairly typical fate for older physicists, as we have seen with Infeld and Rosen. During the 1970s, he worked in the UK Ministry of Defence and was much less involved in scientific work. Thus, from about 1965, he ceased to play an active role in gravitational wave theory.

As we have seen, none of these three skeptics seemed to impress their concerns deeply on their students and collaborators. Bondi's uncertainty about whether freely gravitating dust would radiate remained largely private. Another worry of his was the existence of tails in gravitational waves. In 1966 he described this as an "absolutely disastrous discovery," which might indeed lead to deep insights but was nonetheless "extremely serious," in that it prevented a perturbed system from ever entirely settling down to a static state again, because its field would always be affected by its own once-turbulent past history (Bondi 1970, p. 269; p. 270). Certainly, tails have remained an important topic in the study of gravitational waves ever since, and some have even suggested that they might interfere with the detection of gravitational waves, but on the whole, posterity has not shared Bondi's concern, which he still feels should be addressed (interview). He perhaps recognized the essentially accommodating instinct of most physicists when remarking, "No doubt we can live with history in gravitation. But that does not prevent me from regretting [it]" (Bondi 1970, p. 272).

Even if Rosen, Infeld, and Bondi could not always convince their colleagues to focus on the issues they regarded as fundamental, each made great contributions to the subject. Infeld did so through the EIH formalism, his fostering of the post-Newtonian approach to radiation problems, and his mentoring of an army of relativists and theoretical physicists, many of whom worked on problems related to gravitational radiation. Rosen stimulated debate on gravitational waves throughout his provocative career and also contributed through his students, such as Peres. Rosen was such an outspoken iconoclast that his own individual style of physics remains well known even after his death and still has admirers. Bondi, with his various collaborators, did as much as anyone to create a concrete picture of gravitational waves as a real theoretical phenomenon, rather than an abstract mathematical analogy. Perhaps the skeptical contribution is best summed up, if it must be encapsulated simply, in terms of that process. If one thing united the skeptics, it was their resistance to a straightforward imposition of an analogy with electrodynamics onto nonlinear gravitation theory. Complacency with this analogy indeed posed a great threat to the idea of gravitational waves, since if it were accepted too sweepingly, they might never have achieved an independent existence worthy of note and remained nothing more than a footnote to field theory. If for some, such as Pirani, the need to deepen the analogy, by developing a quantum theory of gravity, served as a motivation for the study of gravitational waves, for others the need to question the analogy was just as compelling a motive. By their efforts, the foundation of a theory of gravitational radiation was laid in the 1950s, just in time for the historical moment of renewed interest in general relativity that followed in the 1960s and that witnessed interest in gravitational waves reach beyond the narrow borders of theory for the first time.

If the ideas of the skeptics were taken up in quite a different style by a new generation, it is perhaps the general fate of scientists to pass on to posterity not their whole idea, but just an element of it, which then becomes a tool for their successors to use in constructing their own concepts. Indeed, if there can be some regret for older scientists in the failure of younger generations to properly understand their work and motivations, this repression of the personal in the history of physics is perhaps what attracts people to it. Creative young scientists do not feel burdened by the dead weight of the personalities of aging or deceased scientists when attractive elements of their thought can be appropriated, stripped of the sensibilities that originally animated them, and be given an entirely new meaning in a new style of physics. In interviews with both Bondi and Chandra, I was struck by how both of them observed that their successors had not necessarily understood the motivation that lay behind their best work. Chandra, in particular, seemed quite melancholy on this score. Bondi remarked to me that "nobody has fully understood" the news function, his most famous contribution to the study of gravitational waves, but he did not seem unduly bothered by this.

Not only is it difficult in physics (no doubt in any walk of life) to really imbue a field with one's own outlook, but even having students as a means of expanding ones intellectual influence can be a double-edged sword. It seems as if taking on only a few students makes it more likely that their views will reflect their mentor's, but it also means a smaller chance that they will influence the field as a whole. Have more students, and they may develop into an important school and occupy many positions in the field, but as students they may tend to respond to their own peer group and may not agree with all the viewpoints of their mentor. No doubt in such cases it is the students who influence the teacher about as often as the teacher influences the students. After all, they outnumber him or her.

10

On the Verge of Detection

By the time of the jubilee conference at Bern in 1955, gravitational waves had had a theoretical existence within general relativity theory for close to forty years, yet it would be fair to say that no convincing picture of their nature and behavior as a physical effect had yet been formulated. Certainly there had been some debate already as to whether they existed or not, whether they carried energy away from real systems and so on, but in the leading textbook on relativity after the war, Peter Bergmann's *Introduction to the Theory of Relativity* (1942), we find gravitational waves (which are "typical for a field theory") introduced as "rapidly variable fields, which must originate whenever mass points undergo accelerations" (p. 187). This description can hardly have induced a vivid image of the waves in the minds of his readers, and one must suspect that this kind of phrasing is more intended to obscure what is not known than to describe what is known. One cannot help feeling that, in this period, relativists were somewhat embarrassed by this stepchild of field theory, which had been foisted upon them with no clear theoretical role and no prospect of experimental significance and which had proved so intractable to theory.

While we have seen that the abstract style of analogy with electromagnetic field theory played an essential role in encouraging the growth of a theory of gravitational waves, relativists were slow to construct a *descriptive* analogy, or metaphor of the phenomenon. Perhaps this reflected the lack of a compelling theoretical understanding of the waves themselves. In many postwar popular and textbook discussions of gravitational waves, metaphor is completely eschewed in describing the waves, the authors having preferred to substitute an explanation of the effect of the waves on some idealized system, such as a circle of particles, which would be deformed into an elliptical shape by the passage of the wave (e.g., Goldberg 1966). That the need for some metaphor which might strike a chord with the reader was felt is illustrated by the use of a comparison with shear waves, one of two main types of acoustic waves in solid matter, in Bergmann's *The Riddle of Gravitation* (1968). Some authors such as Wheeler (Wheeler 1962) and Feynman, in his lectures on gravitation (Feynman 1995), chose to elaborate the electromagnetic analogy more fully, down to the quantum analogy between "gravitons" and "photons."

By the early 1970s, at least, we find that confidence in the theoretical understanding of the waves had increased to the point where relativists at last felt comfortable in reaching back to the metaphor underlying all descriptions of wave phenomena, that of ripples on water. In the best-known modern textbook, by Misner, Thorne, and Wheeler (1973), we find the use of this physical metaphor right at the beginning of the chapter on gravitational waves. Their version is perhaps the canonical one:

> Just as one identifies as "water waves" small ripples rolling across the ocean, so one gives the name "gravitational waves" to small ripples rolling across spacetime. Ripples of what? Ripples in the shape of the ocean's surface; *ripples in the shape (i.e. curvature) of spacetime* [emphasis added]. Both types of waves are idealizations. One cannot, with infinite accuracy, delineate at any moment which drops of water are in the waves and which are in the underlying ocean: Similarly, one cannot delineate precisely which parts of the spacetime curvature are in the ripples and which are in the cosmological background. But one can almost do so; otherwise one would not speak of "waves"! (pp. 943–944)[1]

In Banesh Hoffmann's 1972 biography of Einstein, the same description is found, "ripples of curvature traveling with the speed of light," along with "frozen corrugations of space-time acquiring for us the aspect of motion because of our passage through time" (pp. 219–220). This shows that by the seventies, the picturesque physical metaphor was thought suitable for inclusion in a popular account.

Different similes continue to be employed side by side. The *McGraw-Hill Encyclopedia of Physics* (Parker 1983, p. 404) uses both "ripples in the curvature of space-time" and "propagating patterns of strain." Rees,

Ruffini, and Wheeler (1974) stated that Einstein showed "that geometry can undulate and carry energy" (p. 84). They later added, when discussing the skeptics (while noting that "doubts [of the reality of gravitational waves] [of] earlier days have now been dissipated") that "any talk of a gravitational wave carrying energy is nonsense. There is no such thing as the local density of gravitational wave energy."

A popular book, *The Search for Gravity Waves* (Davies 1980), begins with the electromagnetic analogy but "caution should be exercised in stretching the analogy too far" (p. 26). It also introduces gravity waves as ripples of geometry but warns against the danger of confusing them with mere coordinate ripples (p. 49).

In contrast, Weber's seminal text on gravity wave detection (1961) does not allude to any analogy other than the abstract electromagnetic one. There is no mention of ripples in spacetime, geometry, or anything else.

If, by 1970, relativists were sufficiently confident in their own theoretical picture of gravitational waves to begin advancing a more compelling metaphorical description to the student or lay person, we may look for the source of this self-assurance in the theoretical work of the previous decade, which laid the foundations for a physical (as opposed to merely formal, mathematical) theory of gravitational waves. Two research groups in particular played a key role in this endeavor: the Bondi-Pirani group in London and the Wheeler group at Princeton.

As we have seen, the Bern conference itself helped to inspire Bondi to take up the problem of gravitational waves. Both he and Pirani were motivated by the uncertain status of the phenomenon at that time to address the issues that had been raised (by McVittie, Rosen, Infeld, and others) as to the existence of gravitational waves. They began, together with Ivor Robinson, by rebutting the long-standing objections to the existence of plane waves in gravity (this view was sufficiently orthodox in the immediate postwar period to enter the textbooks [Bergmann 1942]). This was of considerable importance, since the cylindrical waves, the only other exact wave solution of Einstein's equations available at that time, were quite unphysical, requiring an infinitely long source for generation. The basic situation was further improved with the publication of a solution with spherical wave fronts by Robinson and Trautman (1960). With such exact solutions in hand, the London group could proceed to a rigorous study of the asymptotic behavior of the waves, meaning their behavior far from the source, towards infinity. While still idealized, the systems they would study, involving a source and the distant waves generated by it, would be an acceptable idealization, modeling a compact, isolated source.

Bondi has described his 1962 paper with van der Burg and Metzner as "the best scientific work I have ever done" (Bondi 1990, p. 79), and it certainly played an influential role in persuading people that gravitational waves really could transmit mass and energy away from a source. This is the paper in which Bondi introduced the famous news function. The paper itself gives a clear sense of how the particular concerns that motivate a scientist to his best work can be completely lost on his audience, even when the paper in question is rather successful and influential.

The paper touches on two problems that Bondi regarded as very troubling. The first was the question of whether self-gravitating sources (such as binary systems) could radiate. Although by this time some relativists no longer gave this matter much thought, Bondi still quoted Infeld's *Motion and Relativity* approvingly on the matter of radiation from binary systems:

> The lack of radiation for freely falling particles emerges from Infeld's work, but one would like to generalize this to non-singular equations of state. The most clear-cut case then would seem to be pressure-free dust, but beyond this it is tempting to suggest that perfectly elastic equations of state do not lead to radiation. (Bondi, van der Burg, and Metzner, p. 47)

This is interesting, since Bondi was surely well aware of Trautman's work showing that the radiation terms in EIH were not coordinate independent, which Trautman carried out while he was in London at Bondi's department. Neither Bondi nor Infeld regarded Trautman's 1958 papers as convincing proof that freely falling systems ought to radiate.

Bondi's second worry concerned the presence of infinitely long tails in gravitational radiation. This phenomenon was demonstrated conclusively in the 1960s in the work of Newman and Penrose (see below). Tails are disturbances that fail to "keep up" with the rest of the wave, thus violating Huygens's principle, that the total disturbance is confined to a single expanding wave front defined by the speed of the wave. In general relativity, tails, which arise as the wave interacts with the curved background metric of the source, bounce around indefinitely, preventing the system's field from ever quite settling down to a quiescent, static state again. What is happening here is not that some parts of the wave are traveling more slowly than others, but that, when spacetime is not flat, some parts of the wave are reflected, as off the side of a bowl, and begin heading backwards. If a part of that retrograde wave now gets reflected back again, it will recover its original direction of motion, but some distance in the rear of the expanding wave front of which it was originally a part. This is the "tail" of the wave.

The source will always be surrounded by echoes of the original disturbance that caused the original burst of radiation, however short-lived. Bondi referred to the discovery of tails in gravitational waves as "absolutely disastrous," since it showed "the world to be a much more complicated place than had been thought" (Bondi 1970, p. 269). The problem, as Bondi saw it, was that the source's future behavior now depended "on its earlier history right to the year dot . . . and this dependence on history is something which I think we can now definitely identify in the general theory and which makes it a markedly less attractive theory." He concluded, "we have to live with this theory . . . [but] it shows itself to be a little nastier than might be expected"(p. 271).

When I visited Bondi in 1995, this problem of tails in general relativity was still one that bothered him greatly. In another interview, Ted Newman well recalled Bondi's strong and controversial rejection of tails until the work of Newman and Penrose compelled him to accept their existence. Part of his concern can no doubt be found in a practical consideration from his 1962 paper. In that paper Bondi managed to show that a massive system which is initially and finally static, but which goes through some non-axisymmetric disturbance in between, will have lost mass in the meantime to the gravitational waves it emitted. The period of emission is characterized by a certain function, called the news function by Bondi, which is nonzero only during this period. One reason for his insistence on initial and final staticity is that, in general relativity, the problem of defining the mass of a system is a rather tricky one, especially if the system is dynamic. One way around this problem, increasingly in use from the 1960s on, is to look at the field of the (isolated) system far from the system, where it should approximate to the Schwarzschild metric, which describes the gravitational field of a single body. In this context, one can define the mass as being that of a Schwarzschild body with an equivalent field. Bondi thus laid great stress on the initial and final staticity of his system, even though he was successful in defining a time-varying mass function that reduced to the static-field Schwarzschild mass in the static case. Tails that refused ever to die away completely were one obvious threat to this picture.

On the other hand, a system that consisted of freely falling dust appeared to lack any mechanism for knocking itself out of its initial quiescent state. Such a system posed a threat to the other boundary condition, the initial one. It

> does not contain news in this sense. Its future is a clear consequence of its past, and it would seem difficult to draw a distinguishing line between different systems of this kind though conceivably the pressure-free gas might be the only non-radiative material, all others radiating if in motion (Bondi, van der Burg, and Metzner 1962, p. 48).

Bondi regarded the news function as *demarcating* those systems that could radiate from those that could not. "News" was the characteristic of a radiating system; the possibility that information would be carried by radiation out to a distant observer rather presumed the existence of something worth reporting. For Bondi, the medium was indeed the message, in the sense that no possibility of a message implied no medium (no waves). This distinction was crucial to Bondi's interpretation of the news function, which "nobody has fully understood" in Bondi's view (interview). From 1960 on, only a handful of researchers, mostly from or influenced by the older generation of skeptics, continued to feel that freely falling systems did not (or were likely to turn out not to) radiate. Bondi himself, who ceased to be active in the field from 1970 on, did not convince himself that even the idealized freely falling dust case should radiate, in the form of two dust-filled stars orbiting each other, until many years later (interview). In spite of his own doubts on this score, his work and that of others associated with his and Pirani's group in London played a vital role in developing an increasingly convincing picture of gravitational waves in the 1960s. His skepticism, too, played a role in convincing another celebrated astrophysicist, Chandrasekhar, to take up the back-reaction problem (see chapter 8).

Bondi also gave (in part D of Bondi, Van Der Burg, and Metzner 1962, which section was written by Bondi alone), an important treatment of the reception of a gravitational wave. His receiver consisted of "two massive particles . . . with a motor between them." He analyzed its absorption of incoming waves, consequent motion, and, assuming the waves are weak and therefore a linearized scheme is appropriate, its own emission. He was able, based on his analysis, to derive the quadrupole formula for energy absorption, "identical (apart from a numerical factor) with the electromagnetic one." Unlike the back reaction quadrupole formula, Bondi's quadrupole formula has been generally accepted as correct. The problem for skeptics, like Bondi himself, was whether a nonlinear source, such as a binary system, would emit quadrupole radiation at all or according to the same formula. His formula for the *receiver* shows that the wave very far from the source indeed carries a flux of energy with a quadrupole characteristic. Many physicists regarded it as acceptable to equate this far-zone or receiver, quadrupole, with the near-zone quadrupole moment of the source. Others, however, such as James Anderson, criticized this as "proof by naming" (interview), since they thought there were no grounds for assuming that the two quadrupole formulas described the same quantities.

Interestingly, Bondi was not content to let the analogy with electromagnetic waves go without close investigation at either end. While studying the reception of gravitational waves, he was led to look more closely at the theoretical understanding of the reception of electromagnetic radiation. He noticed that this had only been done on a very idealized, and not terribly realistic, basis and consequently wrote a paper in 1960 analyzing this topic from a less idealized perspective (see Bondi 1987 for an account).

The work of Ted Newman and Roger Penrose alluded to above was one of the leading advances in the study of general relativity in a decade of great advances, the 1960s. Together these two mathematical physicists—one American, one English—developed a subtle and powerful way of deconstructing the distant gravitational field of a mass, so that the complex tensorial equations of general relativity could be reduced to ordinary differential equations, for the solution of which physicists have centuries of experience to aid them. Their method made use of tetrad and spinor calculus and provided a straightforward and unambiguous way of doing the sort of "peeling" away of terms in the "asymptotic" or distant spacetime of a source that lay at the heart of the work of Bondi's group and that of Rainer Sachs, one of their collaborators (Sachs and Newman were students of Peter Bergmann's). Newman and Penrose's formalism revealed the structure of spacetime itself in those distant reaches of space far from any stars or galaxies. Contrary to Einstein's original vision of general relativity, far from any mass there is still something. In the viewpoint of modern relativity, there is the gravitational field itself. In relativity theory this field is nothing less than the underlying geometry of space and time, and it is the structure of this geometry that Newman and Penrose laid bare. Of course the study of the "asymptotic structure of spacetime" arose in a natural way from the study of gravitational radiation, because departures from straightforward asymptotic flatness are generally associated with gravitational radiation. The title of their famous paper (1962) is "An Approach to Gravitational Radiation by a Method of Spin Co-efficients," and the formalism is still an indispensable part of work on gravitational waves and many other aspects of relativity theory today. The work of Newman and Penrose is an excellent example of how it is that divergent viewpoints in physics become convergent through technical advances. Whatever motivations have originally inspired it, whether they were mainstream, minority, or even downright quirky, any research program that matures and begins to generate useful research tools will find that these tools are taken up by others. Once physicists are using the same tools, they begin to find that their results agree more often than not. As long as their results agree, they do not worry so much about matters of interpretation or about such philosophical questions as whether tails are a bad thing or whether idealized binary stars made of pressureless dust can radiate gravitational waves.

Nevertheless, the excavation of the very structure of spacetime itself had philosophical implications. Relativity began as a radical program aiming to establish the relationalist view of spacetime: that neither space and time exist independently but are only perceivable as relations between material objects and the events associated with them (thus we say, the table is situated two meters from the chair, or A sat on the chair two minutes before B sat at the table). The alternative view is that space is nothing less than the preexisting container into which tables and chairs can be put and without which things would have nowhere to exist. Though Einstein's early understanding of his own theory was that spacetime simply did not exist in the very distant reaches of spacetime far from material things, he gave up this view and contented himself with the belief that he had at least shown that spacetime's structure was determined by matter, so that without matter there would be no structure to spacetime. By the early sixties the nature of the structure imposed by matter on distant spacetime was at last beginning to be understood, thanks to the study of gravitational radiation (for readings on the philosophy of space and time, one can begin with Earman 1989 and Stachel 2002).

On a practical note of major significance for our study, the body of work on the asymptotic structure of spacetime that culminated in Newman and Penrose's paper had sharpened ideas concerning boundary conditions in the radiation problem. It turns out that there are many different infinities in spacetime, including spacelike infinity, past timelike infinity, and future null infinity (where gravitational waves actually end up), among others. By the end of the sixties it was finally clear that the vague boundary condition which states that there should be no energy carried into the source from outside systems could be precisely stated as no-ingoing or outgoing-only boundary conditions at the appropriate version of infinity (see figure 10.1).

At Princeton, John Wheeler and his students were also interested in addressing questions concerning the physical description of gravitational waves. Wheeler and his group had some influences acting on them that were unusual. For instance, Weber spent time on a fellowship with Wheeler while he developed his ideas of detecting gravitational waves, and this naturally helped stimulate some interest there (Misner, interview). Of course Wheeler himself participated in the debate over whether cylindrical waves could carry energy, and his students were exposed to the controversy at conferences in the late '50s (such as Chapel Hill). Like Bondi, Wheeler and his group disliked the pseudo-tensor approach to the problem of gravitational mass-energy and preferred to examine the distant field of a source to determine its mass (two important definitions of gravitational mass from this period are the Bondi mass and the Arnowitt-Deser-Misner (ADM) mass, Misner being one of Wheeler's students).

Another topic that focused the Princeton group's attention on gravitational waves was the geon, Wheeler's name for an entity constructed of a wave bundle held together by its own gravitational attraction. Wheeler had examined the idea of an electromagnetic geon, constructed from high-frequency electromagnetic waves, and the idea was partly of interest because of a sense that such a "body" might prove a classical prototype for a model of elementary particles composed of pure field (Brill, interview).

Wheeler was already a distinguished theoretical physicist when he turned his attention to general relativity in the 1950s. There was a sense in which

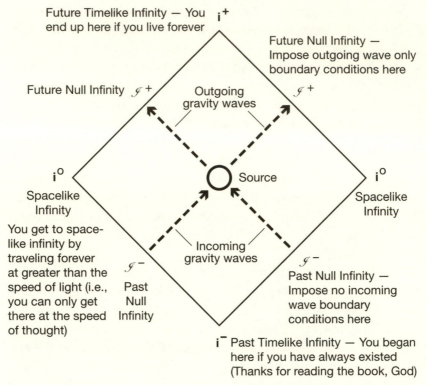

Future Timelike Infinity — You i^+
end up here if you live forever

Future Null Infinity —
Impose outgoing wave only
boundary conditions here

Future Null Infinity $\mathscr{I}^+$ Outgoing
gravity waves $\mathscr{I}^+$

i^o i^o
Spacelike   ○ Source   Spacelike
Infinity   Infinity

You get to space-
like infinity by   Incoming
traveling forever   gravity waves
at greater than the $\mathscr{I}^-$ $\mathscr{I}^-$
speed of light (i.e.,   Past   Past Null Infinity —
you can only get   Null   Impose no incoming
there at the speed   Infinity   wave boundary
of thought)   conditions here

i^- Past Timelike Infinity — You began
here if you have always existed
(Thanks for reading the book, God)

Figure 10.1. Penrose diagram. These diagrams are also known as compactified spacetime diagrams. They use special coordinates that permit infinity to be plotted on an A4 sheet of paper. Gravitational waves, like light, travel along null lines in spacetime diagrams. They head toward Null Infinity. The script letter $\mathscr{I}$, which is used to denote Null Infinity, is called Scri by relativists. Fittingly, this is pronounced the same as the archaic term Scry, meaning "to see or perceive." Timelike infinity is where sub-light speed world lines end up.

gravity was seen as the next stage in the development of quantum field theory. Unlike his student Feynman, who was perhaps more typical in being primarily interested in quantum gravity, on which only slow progress was made at this time, Wheeler adapted effectively to the mores of the relativity community while retaining his own unique approach to physics. He certainly viewed the geometric description of spacetime as an essential feature of the theory, rather than as a curiosity of a wayward and eccentric field theory.

Wheeler set his students Dieter Brill and James Hartle to work on the problem of the gravitational geon. This would be a geon constructed out of short-wavelength gravitational waves, which were especially of interest in that they would be pure sourceless gravitational field constructs, disturbances in the gravitational field held together by their own gravitational attraction. A tool for the investigation of this problem had been borrowed by Wheeler from optics, the idea of the two-lengthscale expansion, or shortwave expansion, in which very strong (and therefore highly nonlinear) gravitational fields could be expanded in terms of the ratio of a short lengthscale describing local disturbances in the field and a long lengthscale describing the background curvature in which the disturbances found themselves. This provided a scheme in which waves could be studied in the context of their own background curvature. That is, the metric was divided into a long-lengthscale, time-averaged curvature representing the gravitational field produced by the waves' mass-energy, and a short-lengthscale, locally varying field representing the actual waves. Brill and Hartle (1964) were able to employ this scheme to study the geon, and Brill (1959) was able to show that toroidal wave pulses appeared to have mass when seen from a distance. (Toroidal wave pulses were first proposed as a spatially limited version of the Einstein-Rosen cylindrical waves by Weber and Wheeler (1957), and a suitable metric was subsequently suggested by Bondi.) The gravitational geon was one way in which the reality of gravitational waves as an energetic phenomenon was argued at this time. Indeed, in recent times Fred Cooperstock (see chapter 11) has attempted to show that the gravitational geon cannot exist. He has made that effort in order to support his energy localization hypothesis, which argues that gravitational field energy cannot propagate through a vacuum.

Subsequently, Richard Isaacson, who was a student of Wheeler's student Charles Misner and who further developed the shortwave expansion description of gravitational waves, was able to produce a tensor quantity, averaged over several wavelengths of a wave, that described the wave energy in an invariant way. This quantity provided a practical means by which the energy in gravitational waves could be calculated for the purposes of estimating their flux of energy and also offered a way to avoid the controversial pseudo-tensor approach to estimating energy in the wave.

The two-lengthscale metric scheme played an important role in visualizing gravitational waves within general relativity theory. The picture of gravitational waves as small-scale "ripples" of curvature (gravitational field) superimposed on the large-scale background curvature of spacetime sparked the introduction of the now commonplace metaphor of gravitational waves as "ripples in the curvature of spacetime" (Thorne, pers. comm.). It was certainly Wheeler, with his gift both for visualization and for neologisms (*geon*,[2] *black hole*), who popularized the spacetime curvature picture of general relativity with his book *Geometrodynamics*, published in 1962. It was Wheeler himself who pointed out the nineteenth-century Riemannian premonition of our contemporary picture of gravitational waves contained in a paper by the mathematician William Clifford (Wheeler 1961). Clifford, a well-known Victorian mathematician who popularized Riemannian geometry for an English-speaking audience, was inspired by Riemann's generalization of non-Euclidean geometry:

> I hold in fact (1) that small portions of space *are* in fact of a nature analogous to little hills on a surface which is on average flat; namely, that the ordinary laws of geometry are not valid in them. (2) That this property of being curved or distorted is continually being passed on from one portion of space to another after the manner of a wave. (3) That this variation of the curvature of space is what really happens in that phenomenon which we call the *motion of matter*, whether ponderable or ethereal. (4) That in the physical world nothing else takes place but this variation, subject (possibly) to the law of continuity. (Clifford 1876, quoted in Wheeler 1961, p. 63)

Clifford's idea of *curvature waves* was quite different in orgin from Poincaré's idea of the *wave of acceleration*. Einstein's gravitation theory, which married Riemannian geometry and relativistic field theory contained the possibility within itself of uniting the two pictures, but it was not until the 1950s and '60s that they were really unified in the modern depiction of gravitational waves. Wheeler's vision of the versatility of the gravitational field, supporting large-scale curvature and small-scale ripples, found expression in the two-lengthscale approach and undoubtedly helped stimulate a new visualization of the idea of gravitational waves. Brill himself recalled that "the idea of small scale ripples was around" at that time (interview), although it may have been some time before the physical metaphor employed in Misner, Thorne, and Wheeler really gained currency.

What is the role of the theorists today? Their great controversies have subsided, the experimenters are working on their instruments, and it might appear that the theorists have little better to do than wait for the results to come in. Nothing could be further from the truth. The job of the experimenters is fantastically hard. LIGO must measure an extremely small shift, about the width of the nucleus of an atom, in the position of its mirrors (which are 4 km apart) to successfully detect gravitational waves. Not surprisingly the motion of the mirrors caused by various other effects (what is called "noise" in the detector) will be just as great. In order to extract the signal from the noise, it is necessary to know in advance what the signal looks like. It is just as if one were trying to find one's way on a foggy night; knowing which landmarks to look for would greatly help in finding them. Because the signals have never before been seen, only theorists can provide the signal templates that will be needed for the sophisticated data analysis these detectors will require. Thanks to the work described in this book, a very detailed picture has been developed of the evolution of compact binaries and the gravitational waves they emit. But it is all based on approximations of Einstein's theory that break down completely as the two bodies actually approach each other and merge. This is the step, unfortunately, during which the strongest burst of waves is emitted. It would be nice to have an exact solution of this final stage of the binary inspiral and merger. To get that, supercomputers that perform prodigious feats of calculation are required. For many years already, dozens of physicists have been working in small and large groups to try to solve the exact Einstein equations for the problem of motion. One of the leading theorists, Kip Thorne, has even made a bet with the leading numerical relativists (as the computer-aided theorists are called) that they will not have solved their calculational problem before the experimenters have actually seen the first colliding black holes (aided by the approximate calculations that Thorne and other theorists specialize in). It remains to be seen who will win the bet.

The theorists, despite the difficulty of their own project, can take comfort in the exceptional technical difficulty of detecting gravitational waves. Although gravitational waves at first featured in the burst of new ideas emerging in the excitement surrounding quasars, pulsars, and other discoveries in astrophysics in the fifties and sixties, they were soon left behind as researchers turned to hotter topics with more obvious astrophysical applications. The subject of gravitational radiation still seemed handicapped by the lack of any experimental input. Although in the late sixties theorists grew more confident about the existence of gravitational waves and there was more optimism than had prevailed previously regarding the possibility of observational results, the actual experimental evidence was still quite meager—and all negative.

In the early 1970s, as if on queue, the actual detection of gravitational waves emerged as a heated and controversial topic for the first time. If this was unexpected, the subsequent emergence of an apparently ideal test bed for the observation of gravitational radiation reaction effects was even less so. By the 1980s radiation reaction in general relativity was a flourishing subject, and the advent of ambitiously large detector projects was creating a demand for unheard of and previously undreamt of levels of precision in the theoretical prediction of gravitational radiation effects. Much sooner than anyone might have expected, the visualization of gravitational waves as a concrete, physical phenomenon, which was the principle achievement of the theoretical advances of the late 1950s and 1960s, was bearing fruit in the form of attempts to render the phenomenon visible via instrumentation. It was only in keeping with the history of the subject that this endeavor would prove to be the most controversial yet.

As we have seen, great progress had been made during the sixties on many fronts in the description of wave propagation and interaction with matter. Supernovae and binary neutron stars were suggested as possible astrophysical sources during this period, as inspired initially by Joseph Weber's work (Dyson 1963). When Weber embarked on his experimental program around 1960, there was no existing theory of sources in the practical sense. Most theoretical work on gravitational wave emission had focused on binary systems, with the Earth's own orbit around the Sun frequently given as an example. Those who were most convinced of the reality of gravitational wave emission from such systems (e.g., Landau, Lifshitz, and Fock) were also most apt to observe that the effect was practically negligible. From Weber's point of view, the challenge was simply to achieve the maximum sensitivity possible with his instrument, and therefore, in that sense, the theory deficit was not critical. However, his chosen design, a resonant bar which would "ring" in response to gravitational waves oscillating at the bar's fundamental frequency, had a rather narrow frequency bandwidth. Therefore it was worthwhile to choose a frequency at which one had the best chance of hearing any cosmic sources that might be out there.

Figure 10.2. Joseph Weber seen with one of his resonant bar gravitational wave detectors. Despite the controversy that enveloped him, Weber insisted to the end of his life that his devices had seen something which was most likely gravitational radiation. (Courtesy AIP Emilio Segré Visual Archives)

Again, the demands of sensitivity restricted the choice at hand. A large bar was necessary to achieve higher sensitivities, but logistical considerations limited the total size. As total size was a principal determinant of resonant frequency, the range of operating frequencies was limited. Weber chose an operating frequency of 1661 Hz, relying largely on his intuition of the subject, which had been built up over a considerable time spent in theoretical preparation, partly in collaboration with John Wheeler, with whom he spent a postdoctoral year at Princeton.

It was only after the beginning of his experimental program that Weber began to receive some suggestions as to possible sources. On one of Weber's subsequent visits to Princeton, Freeman Dyson suggested asymmetric supernova collapse, in which a bump in a star undergoing gravitational collapse would be spun around more and more rapidly as the star shrank, releasing increasing quantities of quadrupole radiation and sweeping in frequency up through the kilohertz range of the bar (Weber, interview). During the collapse of the core of the star during a supernova, the rate of spin reaches enormous speeds. Just as an ice-skater spins faster after pulling her arms in close to the body, a more radially compact body must spin much faster to conserve its total angular momentum. Unfortunately, the quadrupole formula predicts that a perfectly spherical star will generate no gravitational waves, no matter how fast it spins. This was one aspect of the theory upon which everyone agreed. But if there was a bump on the surface of the star, it would cause a changing quadrupole moment as the star spun around, and the faster it went during the collapse, the more waves it would generate. If the bump were big enough, quite a lot of energy could be radiated away as gravitational waves before the star had finished its collapse.

For many years, this type of source remained a favored candidate for gravitational wave detection. It was eventually superseded, however, by another suggestion of Dyson's published in the book *Interstellar Communication* in an article titled "Gravitational Machines" (Dyson 1963). The article discusses possible uses of gravitational energy by "advanced civilizations." One of these is the deliberate creation of binary neutron star systems whose intense gravitational fields and swift orbital motion would be used to gravitationally accelerate spacecraft to enormous speeds (like the gravitational slingshot effect used to accelerate the Voyager spacecraft of the 1970s and 1980s). Dyson observed that "if a close binary system could ever be formed from a pair of neutron stars" (whose individual existence, he noted, is "uncertain") (p. 119), these systems would emit sufficient quantities of gravitational radiation (on account of the intense fields produced at short range by such highly condensed bodies) to cause the system to decay on a relatively short timescale, until its two components plunged into each other in a final immensely strong burst of gravitational waves at a frequency suitable for detection by Weber's instrument. He estimated that such an emission "should be detectable with Weber's existing equipment at a distance of the order of 100 mega parsecs," a pretty good estimate by today's standards. Since this gives a range covering an expanse of space containing up to 10 million galaxies, "[i]t would seem worthwhile to maintain a watch for events of this kind, using Weber's equipment or some suitable modification of it" (p. 119).

At first, Dyson's putative source may have been seen as somewhat science-fictional. He certainly put it forward in an unusual setting, a book about communication with extraterrestrials (a very earnest and unsensational one, to be sure). Furthermore, at the Warsaw conference, we find someone, an "unidentified questioner," apparently taunting Weber after his presentation with a question as to whether he had yet "measured any Dyson neutron binaries" (Infeld 1964, p. 82). This type of source must have seemed almost wildly speculative at a time when the existence of neutron stars was very much in doubt (pulsars were not discovered until 1967 by Jocelyn Bell and Tony Hewish). Nevertheless, pulsars have since been discovered with binary companions (see below), and nowadays, along with the as yet undiscovered binary black hole systems, they are the most favored source for the next generation of gravitational wave detectors.

So the originator of the experimental effort to detect gravitational waves was himself present at and played a role in the debate over whether the theory even predicted their existence in any meaningful form. Throughout the 1960s, Weber ploughed a lonely furrow in his efforts at gravitational wave detection. By the time of the Warsaw conference in 1962, he had a detector operating, and he presented his early results. His reception at this stage was mixed at best. As Weber later remarked, he was laughed at before he had inspired the field that followed in his footsteps, as well as afterwards.[3] Nevertheless, he succeeded not only in constructing a working instrument but also in elaborating some of the most important future developments of the field of which he was as yet the only exponent. His chosen detector consisted of a large aluminum bar seismically isolated from the vibrations of its surroundings and fitted with electronic strain gauges that would detect any resonances set up in the bar by a transient disturbance. He also sketched the idea of an interferometric detector, which is how today's big detectors, such as LIGO, operate. One of his students, Robert Forward (now a well-known science-fiction writer) was the first to build and operate such a device in California in the early 1970s (Thorne 1989). Weber noted that the Earth itself was a large gravitational wave detector on the Weber-bar model, and it set an upper limit on the quantity of gravitational waves passing through it that might excite vibrations in its mass. As the decade progressed, he improved his detector (situated in College Park, Maryland) and eventually set up a second one in Chicago. This arrangement would allow him to eliminate merely local vibrations affecting one detector by looking for coincident disturbances in both. If both detectors, isolated within their vacuum chambers, were disturbed significantly, it might be evidence that a gravitational wave had passed by the Earth.

In 1969 Weber announced that he was detecting pulses in his instrument in excess of what was expected statistically from purely random (Gaussian) noise. Over the next couple of years he produced an increasing volume of coincidence data between his two detectors, including indications of what was called a sidereal correlation. This was an excess of coincidences that peaked at certain times of the day and that varied during the year in such a way to suggest that the source or sources of the waves lay outside the solar system. Such increased sophistication in Weber's claims overcame an initially lukewarm reaction and persuaded several other experimentalists to build and operate detectors on a similar basis. Their eventual failure to see sufficiently convincing indication of gravitational waves interacting with their instruments led to a bitter and protracted controversy with Weber, who continued to pursue his own research on the basis that at least some of his events indicated the presence of gravitational waves. His early results have nevertheless been almost universally rejected by other gravitational wave experimentalists. (For a full account of this fascinating history, see Collins 2004.)

One blow to Weber at an early stage was that he initially claimed that he was seeing a peak of events once every twenty-four hours, which suggested that it corresponded to some celestial object being overhead at that time. But gravitational waves, unlike electromagnetic waves, do not care about the Earth being in the way. They pass right through the Earth with little attenuation, one of the reasons they are so difficult to detect. Therefore they should be just as visible when the object is "underfoot" as when it is "overhead," which suggests a twelve-hour cycle. When this was pointed out, Weber claimed that his data was supportive of such a twice-a-day cycle. But later some physicists began to see this early slip as part of a pattern in which Weber worked over his data so much that he could eventually get it to fit any desired form.

The reaction of general relativity theorists to Weber's findings was somewhat mixed. Although the discovery of gravitational waves might be thought of as good news for theorists, Weber's claims violated all theoretical expectations of signal strength, even though these expectations had been radically transformed by other forces in the decade since Weber began his detection program. On the one hand, Weber's estimated sensitivity (which was and has been the subject of considerable controversy itself) and claimed detection rate indicated an unexpectedly strong flux of radiation impinging on the Earth. The sidereal correlation was taken to indicate a source in the direction of the center of the galaxy, a plausible source given its relatively high density of matter. If the center of the galaxy was the source, and if it emitted radiation isotropically so that Weber's bar was seeing only a small fraction of the total output, then it could be estimated that this region of the galaxy must be losing hundreds of solar masses a year to the emission of radiation—a staggering figure. Such a rate of loss would indicate that the galaxy would disappear altogether on a timescale much shorter than its own estimated age!

Nevertheless, the theoretical reaction to Weber's findings was not simple rejection. There were suggestions that some of the many underlying assumptions behind this calculation might not hold true. It was even speculated that there might be experimental evidence for such a rate of mass loss from the center of the galaxy, in the form of stellar motions in the solar neighborhood (Goldberg 1974).

If a source near the center of the galaxy radiated roughly equally in all directions, as demanded by the quadrupole formula, then the quantity that reached Weber's detectors was only the tip of a very large iceberg indeed. But theorists can show great flexibility when motivated, and some theorists adopted the reasonable attitude that it was up to them to find scenarios that might explain Weber's results. Charles Misner, one of the

leading relativity theorists and a colleague of Weber's at the University of
Maryland, pointed out that if the sources were beaming radiation so that
almost all the energy was directed in or close to the galactic plane, then
the energy paradox would be resolved. Beaming is a well-known feature of
electromagnetic radiation, associated with extremely high order multipoles,
well beyond the quadrupole. Particles moving near the speed of light radiate
electromagnetic waves in a beam, ahead of them, rather like the headlights
of a car. Little radiation is emitted in other directions. But beaming would
be associated with extremely high source velocities (so that the higher-order
terms in the expansion would be important). The idea was that compact
binary systems, like two black holes, would reach rapid velocities only if
they were orbiting very close to each other, and then the elevated rate of
gravitational wave emission would quickly cause the two black holes to draw
even closer and crash into each other. How would these high velocities be
realized and sustained?

Everyone agreed that this kind of scenario would be an occasion when
the quadrupole formula would no longer be the leading order effect. Higher
multipole radiation would take over. Misner realized that it might be possible to tap the huge amount of energy locked up in the rotation of a rapidly
spinning black hole (an idea previously put forward, in the context of gravitational waves, by the famous Russian physicist Yakov B. Zeldovich). It had
been shown that when objects fall in towards rapidly spinning black holes,
if part of the object falls in and part flies back out, the part that comes back
can fly off with more kinetic energy than the whole object had going in
(Penrose 1969). The energy comes at the expense of the black hole's spin,
which slows down. The same thing can happen with gravitational waves,
with some being absorbed by the black hole and the rest being scattered
back out with amplified energy. A black hole can be hit with a burst of
gravitational radiation and reflect a bigger burst back out. In principle,
that reflected burst, known as superradiance, would strike a body, giving
it more energy, just as those spurious incoming waves in the calculations
of Peres and Hu made the stars spiral outwards. If the energy gained from
the superradiant scattering off the black hole was greater than the energy
radiated into space, then the orbiting stars or black hole would not spiral
inward. It would be able to orbit for a considerable period radiating away
strong gravitational waves at a constant frequency. The orbiting bodies
would have to be very close to each other, and thus would be moving very

quickly, and the radiation might be beamed, in which case Weber's device might be tuned to just the right frequency to catch one or more systems with these "floating orbits" (see Misner 1972 and Press and Teukolsky 1972).

It is interesting to note that the superradiance effect just described can also be seen as an instance of tidal friction, such as operates in the Earth-Moon system. The body orbiting the black hole distorts the shape of the event horizon (the surface inside which even light cannot escape from the gravitational pull of the black hole) as if it was raising tidal bulges in the black hole. The black hole spin, creating a kind of tidal lag, introduces a torque that transfers angular momentum from the black hole to the orbiting object, so that the object can spiral outwards, as the Moon does (see figure 10.3). Note that neither of these views (the gravitational wave superradiance picture, or the tidal coupling picture) is more "correct" than the other. In the near zone it is not possible to discern any difference between dynamical tidal effects and gravitational wave effects; either description is equally valid.

This should serve as a useful reminder that the nonskeptic physicists did not just stand for a sterile status quo. They would have been delighted to find evidence for the cases where the quadrupole formula failed, but their pessimism regarding the existence of such cases was based on their experience in similar cases in astrophysics and electromagnetic field theory. They had good reason to expect that the quadrupole would be the lowest-order multipole emission, and the analogy with electromagnetism suggested that higher-order multipoles would dominate only in specific scenarios involving sources with very rapid motions. It was hard to visualize how binary stars, or any stellar mass system, with such motions would last very long as sources. As Dyson had pointed out at an early stage, parking your neutron stars too close to each other would quickly result in them spiraling into each other. However, when an experimental reason arose to look for a source of this type, physicists like Misner were quick to rise to the challenge. Indeed, Misner's 1972 paper is a masterpiece of imaginative thinking in physics, as well as a good illustration of the self-discipline that is essential to true creativity, since he knocks over most of his own proposals one by one. As one of my colleagues put it, noting that even the upholders of the status quo love it when genuinely new physics shows up: "Gravitational-wave studies seem to be prone to 'new effects' and controversies. But the conventional wisdom seems to prevail. (Boring!)" (Alan Wiseman, pers. comm., transparency).

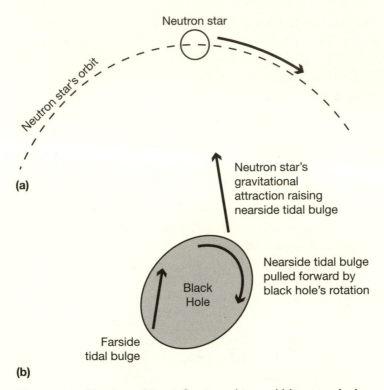

Neutron star

Neutron star's orbit

(a)

Neutron star's
gravitational
attraction raising
nearside tidal bulge

Black
Hole

Nearside tidal bulge
pulled forward by
black hole's rotation

Farside
tidal bulge

(b)

Figure 10.3. Floating orbits. A floating orbit would be created when
the radiation reaction force due to gravitational wave emission, which
is pushing the orbiting neutron star inward, is precisely balanced by the
tidal friction force (or tidal coupling force), similar to that operating
in the Earth-Moon system, which is tending to push the neutron star
outward. The inspiral of the neutron star stops while gravitational
waves continue to be emitted, taking energy from the spinning motion
of the black hole. A black hole takes on an ellipsoidal shape due to tidal
bulges induced on its event horizon by an orbiting neutron star. The
black hole's rotational motion causes the tidal bulges to move forward
of the neutron star's actual position.

Despite these early efforts, no theoretical examples of a floating orbit
would ever be constructed, then or later. It seems that even when an object
draws very close to a large black hole, the amount of radiation scattering
back off the hole can never equal the amount being lost to the universe at
large.[4] In the end, for reasons largely unrelated to theory, Weber's results
were discounted by other experimenters and the community at large.

In its early phase, theorists had only a minor role to play in the Weber controversy. It was not until the late 1980s, when Weber claimed to have detected gravitational waves from the large supernova 1987a under circumstances in which only one other detector was on the air, that theorists came to the fore in rebutting his claims. In that case, experimentalists were ill placed to do so, since the detection was inherently irreproducible (the supernova in question was the strongest seen from earth since the early seventeenth century), and it was accompanied by a new detector theory of Weber's that claimed for his instruments a far greater sensitivity than his original estimate, which was the commonly accepted one. In this way, Weber hoped to circumvent the earlier theoretical objection that there was not enough energy in the galaxy to power his sources. If his detectors were more sensitive than had been thought, that objection lost much of its force.

Nevertheless, the Weber controversy of the 1970s did encourage some increased theoretical activity, and more importantly, by jump-starting the development of an active experimental field of gravitational wave detection, its effects on theory were incalculable in the long run. The very fact that Weber's problems included his violent conflict with theoretical predictions indicated that eventually experimentalists would find themselves dependent on theoretical guidance. Once the early (pre-1975) experiments negated Weber's results, and the decision was made by several groups to persist in the field, they were faced with the necessity of gearing their program to meeting a goal laid down by the expectations of theory. Indeed, the third generation of detectors now operating (such as LIGs) must rely on detailed theoretical predictions of waveforms to make the signal visible in the detector output by the use of sophisticated signal-filtering methods designed to seek out certain patterns that would otherwise be lost in the detector noise.

The next great step for gravity waves came with the discovery of the first pulsar observed in a binary star system in 1974. This discovery, for which Russell Hulse and Joseph Taylor later won the Nobel prize, was immediately recognized as providing the first strong-field observational test-bed for the theory of general relativity. Up until this time, all tests of the theory had taken place in the realm of first-order corrections to the Newtonian theory. However, initial reactions from theorists were pessimistic that the new system would ever show measurable signs of orbital decay due to gravitational wave emission (Damour and Ruffini 1974).

Figure 10.4. Joseph Taylor, who led the radio astronomy team that discovered the first binary pulsar and successfully exploited it as the most powerful laboratory for testing Einstein's relativity theory. (Courtesy AIP Emilio Segrè Visual Archives, W. G. Meggers Gallery of Nobel Laureates. Photo by Robert P. Matthews)

In 1978, after several years' worth of observations on the system, Taylor and coworkers announced that there was definite evidence of such secular orbital decay. Two hundred years after Laplace had first conceived of such an effect, a system was found that perhaps really did exhibit orbital decay due to a retarded attractive force. The announcement was first made publicly at the Texas symposium in Munich, continuing a tradition of important new findings emerging first at one of the meetings in that series. New astronomical discoveries tend to disperse quickly through channels such as International Astronomical Union circulars and by word of mouth (nowadays this occurs even faster via the Internet). For instance, Damour and Ruffini's theoretical paper on the new binary pulsar was published in late 1974, although the discovery paper itself did not appear in a refereed journal until 1975.

It is interesting to consider how neutron star binaries form, since they seemed likely to be an alien engineering work to Dyson. The components of PSR 1913 + 16 orbit so close to each other that if they were still normal stars they would crash into each other. Only their compact size as neutron stars permits their close approach, which is the source of the high orbital velocities and thus their strong gravitational wave emission. But how did they get into such a close orbit, given that they must once have been main sequence stars? The answer is that one of the two stars was already a neutron star when the other star came to the end of its life. The second star would have swelled up enormously in a giant phase, which the Sun will also do one day and become so large it may swallow the Earth. In PSR 1913 + 16, the neutron star must have passed through the outer atmosphere of its companion, accreting some mass but being slowed down by friction, which caused it to descend lower into the atmosphere, so that the orbits were very close by the time the second star collapsed to a neutron star also. Once the two bodies were this close, gravitational radiation reaction could take over and begin to make them decay further. Had they remained at their original orbital radius, gravitational waves would have required longer than the lifetime of the universe to make their orbits decay appreciably.

Although PSR 1913 + 16 is a very rapidly decaying binary by cosmic standards, the pace of its inspiral is still excruciatingly slow by human standards. The length of time over which PSR 1913 + 16 had to be carefully observed to produce evidence of orbital decay gave a slow-motion quality to this confirmation of the quadrupole formula that is quite interesting. The first definitive journal paper on the orbital decay (Weisberg and Taylor 1981) appeared quite some time after the Texas meeting announcement. Even then, caution was still the order of the day, and it was only gradually during the 1980s that exhaustive observational and theoretical work convinced the great majority of experts that the effect was real and not explicable by other influences on the system that had nothing to do with radiation damping, such as a third body in the system, mass loss from one of the stars, some other form of dissipation, and so

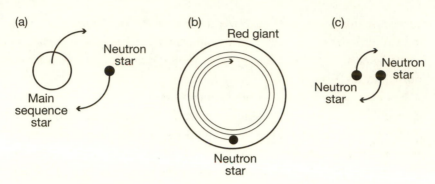

Figure 10.5. Possible creation of a neutron star binary. (a) Neutron star orbiting a main sequence star at a distance of less than 1 AU. The decay in the orbit due to gravitational radiation reaction is negligible. (b) The main sequence star becomes a red giant. The neutron star is within the other atmosphere of the giant, and friction with the gaseous material there causes its orbit to shrink as it losses momentum. (c) The red giant has collapsed to become a neutron star, leaving a neutron star binary with the components orbiting so closely together that gravitational radiation reaction will cause them to spiral into each other within a few million years.

on. Even today, it cannot be logically ruled out that some other unsuspected effect would invalidate the decay's agreement with the quadrupole formula.

Other binary pulsar systems have been discovered since the first one. Observations of the second system for which the orbital decay was measurable did not in fact agree with the quadrupole formula. The reason is believed to be that this system is accelerating with respect to the Earth, so that the rate of its orbital clock is not constant as seen by us. In this case, rather than leading to a renewal of the controversy, the quadrupole formula value was in fact assumed in order to permit an estimate of the galactic position of this binary pulsar, based on its apparent acceleration. This confidence in the quadrupole formula was amply justified by the most recent discovery of the first "double pulsar" by a team at the Jodrell Bank Observatory near Manchester, England. In the double pulsar both pulsars are visible, and the two bodies are closer together than in any previously observed system. In this case the orbital decay is so large that its orbit shrinks by 7 millimeters every day,

(a)

(b)

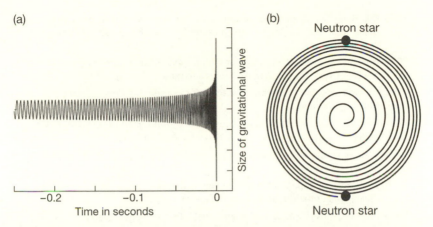

Neutron star

Size of gravitational wave

−0.2   −0.1   0

Time in seconds

Neutron star

Figure 10.6. Gravitational waveform from a neutron star binary. (a) The gravitational wave emitted by the binary has two cycles for every orbit of the neutron stars. As the orbit shrinks, the frequency of the wave increases, slowly at first, then rapidly, as the increasing amplitude of the wave indicates that more energy is being carried away with each cycle. The frequency reaches the kilohertz range before the two neutron stars merge into each other. (b) Since more energy and angular momentum are carried away by the gravitational waves as the binary orbit decays, the rate at which the orbit decays is constantly increasing. Thus the waveform emitted as the neutron stars spiral into each other ends in a characteristic "chirp," a signal with rapidly increasing frequency and amplitude, rather like a birdcall.

and the two bodies will collide within only 85 million years. The rate of orbital decay for this system is in perfect agreement with the theoretical predictions, including that of the quadrupole formula. Thus, a century after the term *gravitational wave* first came into regular use, there seems to be every reason to believe that their existence is beyond doubt.

Not surprisingly, the ferment of the seventies gave rise to hopes that the era of direct detection of gravitational waves was at hand. Kip Thorne's earliest bet on this topic dates to 1978, and in it he wagered the Italian relativist Bruno Bertotti that gravitational waves would be detected within a decade. By 1992 he had "conceded with sad regret." Bertotti "accepted, also with regret," while the referee of the wager, Carlton Caves, a former student of Thorne's, added, with mordant wit, "witnessed with regret" (from the

1970 Bill Burke concedes his wager with his advisor Kip Thorne that the quadrupole formula might not be correct, at leading order, for binary star systems.

1973 Peter Havas objects to the viewpoint of those (like Thorne) who regard the applicability of the quadrupole formula to binary stars as settled.

1974 Arnold Rosenblum, Havas' student, joins Jurgen Ehler's research group in Munich, Germany.

1974 Fred Cooperstock begins his research program based on a binary system held apart initially by a rigid strut between the two stars.

1974 Russell Hulse and Joseph Taylor discover the first binary pulsar, PSR 1913 + 16.

1976 Ehler, Rosenblum, Goldberg, and Havas dispute the validity of the various derivations of the quadrupole formula for binary stars.

1976 Ehler organizers Enrico Fermi school at Varenna, Italy, on Isolated Gravitational Systems.

1978 Taylor and collaborators announce that the orbital decay of PSR 1913 + 16 is observed to be in agreement with the prediction of the quadrupole formula.

1980 Martin Walker and Cillford Will propose their three iterations test of the validity of quadrupole formula derivations.

1980 James Anderson publishes new derivation of the quadrupole formula based on EIH, matched asymptotic expansions (from Burke) and other methods.

1981 Rosenblum publishes fast-motion scattering calculation disagreeing with the quadrupole formula.

1982 Cooperstock and Hobill argue against Walker and Will's thesis that the history of the controversy is essentially settled.

1982 NATO Advanced Study Institute on Gravitational Radiation at Les Houches, France.

1983 Damour's verdict on the binary pulsar data's agreement with theory and the derivation of the quadrupole formula.

1985 From this time on the quadrupole formula controversy is effectively over. For instance, Rosenblum and Cooperstock's last publications in *Physical Review D* on this controversy date from 1984 and 1986, respectively.

1991 Death of Arnold Rosenblum.

1992 Cooperstock unsuccessfully attempts to reopen the debate on the existence of gravitational waves with his energy localization hypothesis.

Figure 10.7. The quadrupole formula controversy.

text of the wager, which is passed on the wall outside Thorne's office at Caltech). (Note that a referee was thought necessary given the likelihood of a disputed detection.) The field had been full of many regrets, many disappointments, and much controversy by this time. Impressively the theorists and the experimenters were filled with a steely determination to carry on for as long as it took to achieve success.

11

The Quadrupole
Formula Controversy

The quadrupole formula controversy of the 1970s and '80s can, in some sense, be summed up by the remarks Feynman made at Chapel Hill in 1957. His words are interesting as a statement of the "non-rigorous" approach to relativity (and theoretical physics as a whole):

> There exists, however, one serious difficulty [in the study of relativistic gravity] and that is the lack of experiments. Furthermore, we are not going to get any experiments, so we have to take a viewpoint of how to deal with problems where no experiments are available. There are two choices. The first choice is that of mathematical rigor. People who work in gravitational theory believe that the equations are more difficult than in any other field, and from my viewpoint this is false. If you then ask me to solve the equations I must say I can't solve them in the other fields either. However, one can do an enormous amount by various approximations which are non-rigorous and unproved mathematically, perhaps for the first few years. Historically, the rigorous analysis of whether what one says is true or not comes many years later after the discovery of what is true. And the discovery of what is true is helped by experiments. The attempt at mathematically rigorous solutions without guiding experiments is exactly the reason the subject is difficult, but not the equations.
>
> The second choice of action is to "play games" by intuition and drive on. Take the case of gravitational radiation. Most people think

that it is likely that this radiation is emitted. So, suppose it is, and calculate various things such as scattering of stars, etc., and continue until you reach an inconsistency. Then, go back and find out what is the difficulty. Make up your mind which way it is, and calculate without rigor in an exploratory way. You have nothing to lose; there are no experiments. I think the best viewpoint is to pretend that there are experiments and calculate. In this field since we are not pushed by experiment, we must be pulled by imagination. (DeWitt 1957, p. 150)

Feynman's emphasis on the power of the imagination to find out what is true is reminiscent of the views of William Blake, who wrote of the conflict between the forces of heaven, upholding the primacy of reason, and the forces of hell, whose cause is the liberating power of the imagination. Although we are accustomed to thinking of physics as the citadel of reason, and Blake himself referred to the tyranny of reason over the other mental faculties as "Newton's sleep," comments such as Feynman's remind us that the imaginative faculty plays a key role in the advance of physics. Feynman argues that imagination cannot be too tightly reined in by analysis, lest the creative impulse be stymied. There must be space and time enough for the imaginative faculties to stand a chance of finding the truth, after which that truth can be rationalized.

The cognitive split between mathematics and physics is characteristic of modern theoretical physics and was there at the birth of general relativity. Vladimir Fock recalled that in the Petrograd of his youth there were two professors who debated topics in general relativity, A. A. Friedmann (the inventor of the expanding universe) and V. K. Fredericks:

> I remember the talks of Frederiks and Friedmann clearly. The style of these talks was different: Fredericks deeply understood the physical side of the theory, but did not like the mathemematical computations; Friedmann stressed not physics, but mathematics. He strived for mathematical rigor and attributed great importance to the full and exact formulation of the initial preconditions. The discussions that arose between Frederiks and Friedmann were very interesting. (Translated and quoted in Gorelik 1993, p. 309)

Between 1957 and the early 1970s a great deal of work at all levels of rigor had indeed been done on the subject of gravitational waves. Many relativists had taken the course advocated by Feynman. Others had been more cautious, but no less successful in many cases, in improving understanding of gravitational radiation. The extent of their achievement was such that by the late sixties, Kip Thorne (like Feynman, a student of John Wheeler and decidedly of the progressive, less-rigorous school) could state that the issue of whether gravitational waves and gravitational radiation reaction existed was at last settled and that the stage had been reached at which applications of the theory, such as the quadrupole formula, could be used in astrophysical applications with little fear of error. At this point, the rigorous relativists might well have asked, what role is there for us? Now that the few years Feynman spoke of have passed, the point at which some rigorous cleaning-up of loose ends might be called for, and we are to be declared redundant in any case! The forces of Heaven began to rally themselves.

In a 1986 talk, "Folklore in Relativity and What is Really Known," Jürgen Ehlers spoke for the virtues of proving what was once imagined. He quoted Synge on the advance of science:

> As science advances, it seems to have a sort of scorched earth policy. The advancing army is full of enthusiasm for its advancing into the unknown, and the unknown is always exciting. If it glances back at the territory it has overrun, it sees little but dullness, the dullness of what seems to be completely known, with little prospect of adding to that knowledge by a deeper understanding. (Ehlers 1987, p. 61)

Ehlers continued, "So, let us now turn to the dullness of what seems to be completely known. Perhaps it is neither dull nor, in fact, really known" (p. 62).

Ehler's attempt to draw the attention of the advancing horde away from its headlong forward rush and back towards its own past points up a curious paradox of the progressive school. In their attitude to history, they are inclined to be conservative. Although the history of their field is not uninteresting to them, they are usually hostile to revisionism. For the progressive, the history of his field has a critical, if passive, role to play as the solid foundation from which further advances in the subject may be launched. As we shall see, attempts such as Ehlers's to critique what has been accomplished are answered with counterhistories designed to reinforce received opinion on what is "known" (what Ehlers called "folklore").

The progressives, as Dieter Brill (echoing his mentor Wheeler) put it to me in an interview, are a class of "daring conservatives." It is their conservatism with regard to what is known that justifies and enables their daring progression into the unknown.

An example of the progressive school in action (Brill's "daring conservatives") is given by Phillip C. Peters and Jon Mathews's 1963 paper on binary inspiral. This paper makes use of the quadrupole formula to calculate the pattern of radiation from particles in Keplerian orbits and the evolution of such orbits under radiation reaction. Already their outlook was markedly different from those who stress how superfluous the back-reaction effect was for any description of orbital motion, and even more so from those who doubt its existence. Peters and Mathews, like Feynman, based at Caltech, state their attitude to the gravitational wave controversy at the outset:

> The linearized version of Einstein's general theory of relativity is strikingly similar to classical electromagnetism. In particular, one might expect masses in arbitrary motion to radiate gravitational energy. The question has been raised, however, whether the energy so calculated has any physical meaning. We shall not concern ourselves with this question here; we shall take the point of view that the analogy with electromagnetic theory is a correct one, and energy is actually radiated. (p. 435)

If Peters and Mathews felt obliged to begin with an apology, however defiant, a decade later the confidence of the Caltech viewpoint had increased, emboldened by new techniques such as those developed by Bill Burke. New discoveries in astrophysics, which have already been discussed in previous chapters, had opened up the prospect of finding actual detectable sources of gravitational waves in the cosmos. Thorne and one of his students, Sandor Kovács, in a paper presenting a fast-motion scheme applicable to nonbound but gravitationally interacting sources, such as those producing gravitational *bremsstrahlung* radiation (see below), outlined an ambitious: program:

> [Because] "gravitational-wave astronomy" may be a reality by 1980, . . . [the] Caltech research group has embarked on a new project: We seek (1) to elucidate the realms of validity of the standard wave-generation formulae; (2) to devise new techniques for

calculating gravitational-wave generation with new realms of validity; and (3) to calculate the waves generated by particular models of astrophysical systems. (Thorne and Kovács 1974, p. 245)

The reference to the "realms of validity" of the "standard . . . formulae," such as the quadrupole formula, expresses an outlook totally at variance with that of the skeptics. Thorne and Kovács were seeking out new realms of application for the venerable quadrupole formula and were inclined to loosen rather than tighten the bounds of its applicability:

> The "quadrupole-moment formalism" dates back to Einstein (1918), and has been canonized by Landau and Lifshitz (1951). The derivations of this formalism which we find in the literature are valid only for systems with slow internal motions and weak (but non-negligible) internal gravitational fields. However, a detailed analysis . . . shows that only the slow-motion assumption is needed; the quadrupole-moment formalism is valid for any slow-motion system, regardless of its internal field strengths. (pp. 245–246)

By 1980 the point has been reached where some consolidation was in order. In Thorne's review paper "Multipole Expansions of Gravitational Radiation," the abstract begins, "This paper brings together, into a single unified notation, the multipole formalisms for gravitational radiation which various people have constructed." Some reference to philosophical issues was still in order:

> The reader should be warned that this article and its author do not aspire to the high level of mathematical rigor and elegance that characterize much of mathematical relativity (e.g. Penrose's (1964,68) conformal treatment of null infinity, and the Bondi *et al.* (1962)-Sachs (1964)-Newman and Penrose (1968) treatment of the asymptotic properties of gravitational-wave fields.) Instead, the author seeks a level of rigor that is (i) high enough to give him confidence of the results derived, but also (ii) low enough to permit the treatment of real astrophysical systems in the real, non-asymptotically flat universe. This philosophy shows up most strongly . . . where the concept of "local wave zone" is introduced to permit a separation of the theory of wave generation from wave propagation. That separation sacrifices the elegant rigor of the Bondi-Sachs-Newman-Penrose approach in

order to treat, e.g. sources embedded in galaxies, with neutron stars and black holes nearby and with a distant, inhomogeneous universe that may curve up into closure. (Thorne 1980, p. 301)

Thus the elegance of much of general relativity can be viewed as a potential obstacle to its relevance, at least to astrophysics, and therefore the former must be sacrificed to whatever extent is necessary to achieve that relevance. The method of matched asymptotic expansions provides the means by which the theories of wave generation and wave propagation can be married together. But the sacrifice of rigor, if not of elegance, was not without its critics. Indeed, in the early seventies, Peter Havas did object to efforts such as Thorne's to declare a satisfactory conclusion to the problem of radiation reaction. To Havas, not only was the validity of the quadrupole formula still at issue, but, in fact, whether there was any quadrupole emission of gravitational waves from binary systems at all was still very much open to question. Of course, Havas was aware that his own efforts to attack the problem via the fast-motion approach were still incomplete, but his objection was to the complacency shown by advocates of the slow-motion approach in overlooking what he saw as fundamental problems with the various derivations of the quadrupole formula. To Havas, these methods had been declared a success when they began to consistently recover results that only appeared plausible because of a reliance on "intuitions," which were trained in "the corresponding problem in electrodynamics." Such intuition could tend to mislead in the context of general relativity (Havas 1974, p. 384).

Without the resources of a large research group to draw on to advance his fast-motion research program, Havas nevertheless had one student in the 1970s who very much shared his view as to the importance of the back-reaction issue in the problem of motion and the failings of the slow-motion expansion techniques. Arnold Rosenblum was not only determined to carry on the fast-motion approach by his own efforts, but he also proved an able and effective propagandist for the counteroffensive of rigor, whose day had apparently come, well over a decade after Feynman had advised gravitational wave theorists, "Don't be so rigorous or you will not succeed" (DeWitt 1957, p. 150).

After receiving his Ph.D. with Havas at Temple University in Philadelphia, Rosenblum went to Munich in 1974 to work with the

Figure 11.1. Despite (or because of) coming from a background in the apparently analagous field of electrodynamics, Peter Havas was one of the most tranchant critics of a too facile reliance on the electromagnetic analogy. In particular he was highly skeptical that the quadrupole formula was applicable to binary stars. In the later part of his career he became a distinguished historian of his own field. (Courtesy Eva Havas)

group of the mathematical physicist Jürgen Ehlers. There, his enthusiasm for the subject of gravitational radiation reaction and his trenchant criticisms of its current state encouraged Ehlers to take an interest in the problem. Ehlers certainly came from the "rigorous" tradition of general relativity himself, the tradition that preferred to deal, applied-mathematics style, in terms of theorems and proofs. His style is certainly a long way removed from that of relativistic astrophysics, which lies at the other end of the spectrum in terms of rigor. From a mathematician's standpoint, the body of work on radiation reaction certainly left a lot to be desired.

Like Thorne and Kovács, Rosenblum saw the problem of two masses scattering off each other as better suited, at lower levels of approximation, to the fast-motion approach than the binary problem. In the scattering problem, two massive objects approach each other from a great distance; interact gravitationally, altering their respective paths; and recede to great distances from each other. In analogy with electromagnetism, the type of radiation emitted is called *bremsstrahlung* (German for "braking radiation"). In the seventies he applied fast-motion techniques developed by himself and Havas to this problem. Unlike Thorne and Kovács (1974), who addressed the same problem via a different fast-motion (or postlinear, as they called it) scheme, his result, in the slow-motion limit, did not agree with the quadrupole formula, giving a loss of energy about twice as great (Rosenblum 1981). Rosenblum was thus led to take a very active part in the debate on the validity of the quadrupole formula, which he himself helped to spark, by the fact that his own calculations disagreed with the established result in one important case. At the same time, he continued his efforts to extend the fast-motion scheme to the case of bound orbits in binary systems (Rosenblum 1982).

Ehlers meanwhile took a rather different approach to the problem. Having not himself worked extensively on the problem of motion, he nevertheless kept abreast of it for pedagogical reasons. He found the literature on the subject unclear, preferring Fock's book as the best treatment. Inspired by Rosenblum's interest, he began to encourage work on the subject of radiation reaction in his own group and to invite visitors with an interest in it, also. He also adopted the role of an independent critic of the various methods used in the field, publishing a review paper on the subject in collaboration with Rosenblum, Havas, and Goldberg (Ehlers et al. 1976) and organizing the workshop "Isolated Gravitating Systems in General Relativity" at Varenna, Italy in 1976 in order to foster efforts to overcome the deficiencies he perceived in the subject (Ehlers 1979).

Ehlers's critique of the subject was wide-ranging. He disliked the use of point masses and worked within his group to discover methods of dealing with finite nonrigid bodies (Dixon 1979). He felt that even with some progress in that direction, the post-Newtonian schemes, which led to divergent integrals at higher orders in the expansion, were highly suspect (a criticism that was aimed by some at Chandrasekhar's work which had also used extended bodies). Burke and Thorne's demonstration that the post-Newtonian expansions should only be used in the "near zone," with matching schemes employed to connect the solutions to wave-zone boundary conditions, was sufficiently encouraging for Ehlers to invite Burke to give a series of lectures at the Varenna workshop. However, from a mathematical point of view, Ehlers considered that the matching schemes still lacked rigor, despite subsequent further progress by Thibault Damour, who introduced another "intermediate zone" in which the matching took place.

For Ehlers, a genuine mathematical relativist, the quadrupole formula episode, despite the level of acrimony that surrounded some exchanges at conferences, provided a welcome level of interaction between relativists of different outlooks. Both at conferences, where the mixing of mathematically and observationally inclined people was encouraged by a common interest, and in his group, where visitors such as James Anderson contributed a different perspective, he enjoyed the mutual exposure to different sensibilities. The controversy served, therefore, to improve understanding of a difficult subject as well as to cross-fertilize between different schools of relativity (interview).

Ehlers can perhaps be viewed as the pure skeptic, not only because of his relative disinterest (he had little of his own work invested in the controversy), but also because he was very much a skeptic in the ordinary sense of "a person who doubts, questions or suspends judgment upon matters generally accepted," to paraphrase Webster's dictionary (*Webster's New Collegiate Dictionary*, 8th. ed., 1977). As a mathematician, he was convinced that gravitational waves were real by the existence of the exact solutions describing them; as a physicist, he was persuaded by Bondi's famous thought experiment. The linearized theory was perfectly acceptable for detectors, and he felt that on a "reasonable physical level" the review paper of Walker and Will (1980) cleared up the question of the quadrupole formula's validity. While some of the issues of method remained outstanding (such as the need to define boundary conditions rigorously in curved spacetime, as opposed to asymptotically flat spacetime), he regarded the quadrupole formula as reasonably well justified both experimentally and theoretically. His criticisms therefore did not spring from motives of either immediate personal interest or from doubts of the existence of the phenomenon in question, but rather from a desire to expose accepted ideas to question, where he regarded them as unsoundly held.

As the host of the Texas astrophysics symposium held in Munich in 1978, Ehlers gave an overview of the state of the field at that conference from the point of view of those he described as the "minority of relativity-theorists" who did not share the widely held opinion that "the implications of GR . . . have been deduced satisfactorily . . . [for] the dominant, secular gravitational radiation reaction effects on the orbits" of systems including the famous binary pulsar, important news of which was announced at the same meeting. He took care to emphasize on what grounds his dissent was based:

> The main shortcoming of [these] . . . calculations is, in my opinion, not that they employ approximations which have not been rigorously mathematically justified—that they share with many approximations used in physics—but rather that they: 1) employ *notions* which are *not well defined* in terms of basic concepts of GR, such as "gravitational field energy," "total mass and linear momentum" of a gravitationally bound body interacting with other such bodies, "point particle," "gravitational radiation reaction force," "near zone," "radiation zone"; 2) use *laws* which have not been *established* within GR, such as an "energy balance between radiation and material sources"; 3) depend essentially on *ad hoc assumptions* which not only are *without foundation* within GR itself, but for which there are indications that they may be *incompatible* with the fundamental assumptions of GR or with each other, such as global coordinate conditions, particular global splittings of the metric into a flat background and a "small" perturbation, non-covariant "outgoing radiation conditions," negligibility of various kinds of "small" terms, etc.
>
> It seems to me to be an important challenge to find derivations of observable relativistic effects, particularly structure and radiation effects, of isolated systems which are free of shortcomings, and which are not based on mere analogies, however plausible, with Newton's or Maxwell's theory. Needed are, in particular, approximation methods which have been rigorously justified at least in theories simpler than Einstein's, and which permit if not an error estimate, at least a reasonable guess about error bounds. (Ehlers 1980, pp. 279–280)

It is interesting to see that Ehlers disclaimed any dogmatic objection to the practice of physics on "rigorously mathematical" grounds. He was aware both of the practice of most physicists and of the difficulty of the problem at hand, which resisted efforts to prove closed theorems. Instead, he grounded his objections on a failure by the majority of relativity theorists to consistently conform to the principles of general relativity theory itself. Specifically, he attacked efforts to import into general relativity, concepts typically found in other physical theories, such as field energy, energy balance, and so on. In short, there was an epistemological disagreement between those who wished to advance relativity theory according to the standards current in the rest of theoretical physics, that is, by attempting to discover within general relativity quantities analogous to those, such as total mass and linear momentum, that would be employed in a similar problem in "standard" field theory, and those who prefered to pursue the matter according to the peculiar tradition of general relativity theory itself, by avoiding concepts that had not been established, with due rigor, within that theory's own framework. The debate between these rival outlooks concerned not only the issue of rigor in calculations, but also the desire for more rigorous definitions of key concepts.

There is an important *philosophical* attack on the practices of radiation reaction theory in Ehler's talk, which is made even plainer at a later conference in Stockholm, Sweden (Ehlers 1987):

> Another statement which seems to be generally believed, usually on rather flimsy arguments, is: **Newton's theory of gravity is a "limit" of Einstein's**. To understand this limit relation is important since 1) Newtonian theory successfully explains many gravitational phenomena and ii) the approximation methods on which the comparison of GR with observations is based assume such a limit relation and even use Newtonian concepts such as masses and linear moments of gravitational interacting bodies, which have no meaning in GR. . . .
>
> Note that [this limiting relationship] is *assumed* in post-Newtonian approximation methods. If this assumption were incorrect, the comparison of GR with observations of the solar system, the binary pulsar or cataclysmic binaries would lose its theoretical basis. Unfortunately, there seems to be no hope of answering this question rigorously in the near future. (pp. 68 and 70)

This is certainly a radical attack on the whole basis of much of modern relativity theory, especially on most of its experimental verifications. It is not surprising that those relativists whose professional affiliations were closer to physics than to mathematics would have had a strong interest in rejecting such a sweeping historical reappraisal.

The main thrust of Ehlers's argument proceeded on entirely epistemological grounds, insisting on the elimination of concepts and assumptions that do violence to "the fundamental assumptions of GR." Only towards the end of his remarks at the Texas meeting did he return to an appeal for "rigor," especially in the technical requirement of approximation methods with some form of error control or estimation, generally agreed to be a persistent failing in this field. On the other hand, as Feynman remarked in the letter to Weisskapf mentioned in chapter 7, when it comes to gravitational waves, the expansion quantities are typically so small that there are few problems in physics to which this kind of approximation is better suited.

It is interesting to note how diametrically opposed the two main responses to the advent of experimental data on gravitational radiation were. Ehlers's response, as a relativist, was to turn once more to the fundamentals of the theory: to derive "observable relativistic effects" that are free of the shortcomings of failing to adhere to general relativity's "fundamental assumptions." For those with an astrophysical bent, the tendency had been to turn outward rather than inward, to conform to the demands of working within a broader physics community which was, by and large, uninterested in the traditional preoccupations of general relativity, sometimes to the chagrin of relativists. In an interview with Alan Lightman Roger Penrose remarked:

> People come in from outside, not being experts on general relativity or cosmology particularly, but knowing about particle physics, symmetry breaking ideas and so on, and bringing this expertise into the subject. I think there are very many more of those people than relativists. Locusts would perhaps be the wrong analogy, but there are huge numbers of people and they see an opening into this subject, and they come in and almost take it over. I felt this a bit with supersymmetry. In general relativity, I felt this again with a lot of the people who tried to quantize it. . . . Bringing ideas in from other subjects is fine . . . as long as the particle physicists appreciate the problems of general relativity. I think often they don't. People come in without being aware of the very fundamental problems we have argued over endlessly among general relativists. There are very fundamental difficulties that one has in trying to quantize, and these people just try to sweep them away. (Lightman and Brawer 1990, p. 429)

Certainly many relativists were fiercely protective of their beautiful theory and not always thrilled at the prospect of its submersion in quantum field theory. One of Havas's papers upholding the skeptical position on gravitational waves, given in Paris in 1972 before the quadrupole formula controversy had really gotten underway, provoked the following outburst from André Mercier, who had kickstarted the renaissance of relativity in 1955 when he organized the jubilee conference in Bern:

> Physicists . . . perhaps are too conservative in believing that physics (theoretical) should always be made and interpreted the same way, e.g. by wanting to do within GRG the same as in Electrodynamics. Perhaps, the revolution implied by GR is that it is precisely another way of concerning the physical world. And Einstein's drama was perhaps that he tried all his life to unify gravitation and electricity, believing or suggesting at least that these two phenomena are alike, i.e. both interaction, if electricity is an interaction. I personally am not sure that mass is a kind of charge, I am not sure that physics should consist in assuming a vacuum and putting things in it, that there are free fields and perturbed fields, etc. Unification in Einstein's sense was never a success. Perhaps the interaction proper (electromagnetic, strong and weak) can be unified; I had some argument about it in my talk last Monday; but not with gravitation, which is not an interaction in the same sense.
>
> I could go on like that, calling attention upon the *fundamental* difference between GRG and physics as it is done elsewhere. (Havas 1974, p. 390)

Completely independently from the work of Ehlers, Havas, and Rosenblum, another challenge to the quadrupole formula came from Fred Cooperstock, a Canadian physicist who had been interested in gravitational waves since his graduate student days. In the early seventies he was moved, by discussion with Achilles Papapetrou, to make an attempt to eliminate the problem of tails from back-reaction calculations by removing the past history of a binary system. Because of scattering of the emitted gravitational waves off the source's own background curvature, a given binary system emitting gravitational radiation would be affected by these tails from all of the previous states of its history, as its old waves came back to haunt it. Tails, which had been so bothersome to Bondi, were somewhat problematic to deal with, since technically they altered the no-incoming-wave boundary condition. A source like a binary star would have emitted waves at some earlier period of its history, some of which would have scattered back to impinge upon the source itself. Thus, even if the source was alone in the universe at any given time, there would be incoming radiation at some small level, and worse, this radiation would depend on a solution of the behavior of the source at that earlier time (itself modified by radiation emitted still earlier).

Cooperstock's idea was to imagine a source consisting of two bodies held apart, in a static system, by a rigid strut. The strut would then be broken, and the two bodies allowed to fall towards each other from rest (Cooperstock 1974). Cooperstock's early results showed a much higher rate of emission than would be expected from the quadrupole formula and also led him to criticize the basis for this result. At the same time, in avoiding one issue of principle with his toy model, he had introduced several others just as serious. A rigid strut was not permitted in general relativity (since its local speed of sound would be infinite, even faster than the speed of thought!), and the fluid bodies themselves would have to be held together by some sort of skin. Therefore Cooperstock began to further elaborate his model, in response to countercriticisms from others. This process of challenge and counterchallenge, in which models and calculations must be endlessly modified and repeated in response to an endless string of objections, is probably seen in any sustained controversy of this type. This aspect of controversies forces physicists to look, uncharacteristically as Ehlers pointed out, backwards at their own history.

An important feature of the quadrupole formula controversy in the seventies and eighties was the series of review papers by different authors, each employing the history of the subject to illustrate a particular view of the contemporary state of the field. These papers show that relativists were keenly aware of the history of their field and that they were able to draw lessons from their reading of history which reinforced the points they wished to make (see Galison 1995 for similar behavior among string theorists). The earliest of these papers was that of Ehlers, Rosenblum, Goldberg, and Havas (1976). They argued that previous attempts to deal with the back-reaction problem were all inadequate in one way or another. In response, they advanced an outline of a program that would overcome these past failings. Essentially an attempt to formulate a research program for the subject, their paper was followed by the Enrico Fermi summer school in Varenna organized by Ehlers, whose aim was also to foster new work in the field along more rigorous lines than before (Ehlers 1979).

In 1980 Walker and Will took a very different tack, addressing the problem of irreproducibility that had plagued the subject and had been a particular thorn in the side of the slow-motion calculations, with their widely differing results (Walker and Will 1980). They argued that a basic iterative algorithm, applicable for both fast-motion and slow-motion methods, could be followed to recover the quadrupole formula from reaction calculations. It will be recalled that in problem of motion calculations involving radiation, one is forced to assume some initial path of motion for the source, calculate the radiation arising from that motion, then calculate the effects on the motion of that radiation, and beginning a second iteration, use the new path to recalculate the radiation. Walker and Will presented an analysis of a cross section of well-known calculations, dating back to Hu's paper in 1947 and argued that those who had advanced three steps in the iteration recovered the quadrupole formula, and that others, with fewer steps, did not, except for a couple who found the result with the aid of compensating errors (obviously a dire warning of the perils of "knowing" the result in advance).

In this view of the history of the field, there existed a definite method by which the standard results could be recovered in a reliable way. Indeed, Walker and Will's iterative test, that three iterations of the field equations were required to successfully account for retarded effects in bound orbits, became a benchmark for subsequent research. Since scattering problems, where the bodies are not gravitationally bound, required only two iterations, they were typically preferred by those employing fast-motion calculations, in which the calculational burden grew excessive at the third iteration. In one case Thorne had made use of Burke's radiation reaction potential (discussed at the end of chapter 8) in a specific gauge to derive the quadrupole formula for a binary system in two iterations (Thorne, pers. comm.). In the Misner, Thorne, and Wheeler textbook (1973), he had attempted the same trick in a different gauge where the usual three iterations were needed. However, an unconscious compensating error enabled him to recover the quadrupole formula where he should not have been able to, as Walker and Will pointed out. Although this was exactly the sort of sleepwalking success that the skeptics feared, Walker and Will's paper was reassuring to the field since it seemed to clarify the steps that were necessary to carry a calculation through correctly.

This was in stark contrast to the view expressed by Ehlers et al., which was to advocate a more general prescription, whose outcome was not yet known. Yet another view was put forward by Cooperstock and Hobill in 1982. They refused to set forth a general scheme or advocate a particular result, instead arguing against preconceived notions (Cooperstock and Hobill 1982). Their history, as befitted their standpoint, was more descriptive than prescriptive, celebrating the diversity in the development of the field. Another protagonist with an interest in and excellent knowledge of the field's history was Thibault Damour. His papers were often prefaced with a discussion setting his work in a historical context (for example, Damour 1982). In this role, the object of history was to motivate the new work being presented, and the focus was on the previous failings that were being addressed by the new contributions (see, for instance, Damour 1983). A more active role for the historical literature was found in the account of James Anderson, who returned to the Einstein-Infeld-Hoffmann scheme, complete with its surface integral method, and married it to the matched asymptotic expansions of Burke, with further additions of his own, to produce another influential derivation of the quadrupole formula (Anderson 1987).

A very significant aspect of the debate in the seventies and eighties was the problem of when theory ends. Recall the title of Peter Galison's book *How Experiments End*, which discusses the thorny problem of how an experimenter is to know when to stop looking for errors in the experimental method and accept the current answer as the correct one. It certainly seems that theorists face the same dilemma, both in their individual research and collectively. As we have seen, different authors could look at the same history and give very different answers to this question. One answer might be that theory already has ended, and we really know the answer ("conservative"). Another is, it has just ended now, with this paper, for the issues addressed ("technocratic"). A third is, it will end as soon as the general program we advance is carried through ("Marxist"). A fourth is that it can never end, and it is best that it should not ("anarchist"). Finally there is the view that the answer is hidden in the past, waiting to be extracted and pieced together from the literature ("archaeological"). It is interesting that just as there was agreement on the details of the history (and the debate was largely a historical debate), opinions diverged on the matter of *interpretation*. The lesson of history was different for everyone. This is still the case, but the debate having lost its impetus, the individual perception of history has lost its public relevance once more. The dynamic of the debate is that some level of consensus must be found for the resolution of an existing problem, and yet progress seems to be measured, for many scientists, by the extent to which an issue can be settled, allowing the next problem to be addressed. A field like general relativity has historical memories of the isolation that may be the fate of a discipline which does not progress in this way. Feynman's remarks at Chapel Hill expressed the view of the progressives, when he said, "The second choice of action is to . . . drive on," to "make up your mind [whether gravitational radiation exists] and calculate without rigor in an exploratory way." He concluded with the advice, "Don't be so rigorous or you will not succeed"(DeWitt 1957, p. 150). The contrast in attitude suggested here may explain why the debate tended to become more vitriolic in its last stages, as a consensus developed for many, while some still argued that the matter was unsettled. The fact that experimenters had at last contributed to the debate did not immediately settle the controversy but did provoke efforts to bring it to a close, possibly contributing some acrimony to what had been until then a rather civilized debate.

At the time of the ninth Texas symposium, held December 1978, for the first time outside the United States (in Munich), a vigorous debate on the problem of motion for binary systems was still underway. A workshop at the conference was devoted to the subject, and the report in the proceedings indicates a wide-ranging discussion and a fairly rich strain of new work in the field. The problem was, however, raised to a new level of prominence as the result of the most exciting development of the conference: the announcement of a measurable orbital decay in the binary pulsar by Taylor and coworkers (discussed in chapter 10). The stimulus of what was generally considered to be rather high quality data on the higher-order evolution of the motion of an apparently isolated binary system containing bodies with strong internal fields led to the 1980s being the most prolific of all decades to date for work on various aspects of the radiation reaction problem and the problem of motion. Subsequently, the demands of a projected new series of gravitational wave detectors encouraged the continuation of efforts to calculate to ever higher orders in the expansion parameters.

For the moment, however, the new experimental data from what Taylor described in the symposium proceedings as "an ideal machine for testing gravitation theories," which might have been designed for the purpose, favored the back-to-basics appeal of Ehlers's tendency.

One of the strongest threads running through the history of twentieth-century gravitation theory is the search for decisive tests of rival theories of gravity. If now a test had been found for the prediction of the quadrupole formula, it had suddenly become a matter of some importance that the quadrupole formula be more rigorously shown to be a consequence of general relativity, honoring its "fundamental assumptions." The experimental data is a test of the quadrupole formula. It can only become a test of the theory itself when everyone has agreed that the formula is definitely a consequence of the theory. Note however, that Feynman, for instance, would probably have disagreed with most relativists about how critical it was to have a rigorous derivation of the experimental result from the theory. For him it was enough to be able to successfully calculate a result in agreement with experiment. To quote once again from his remarks at Chapel Hill, "The real challenge is, not to find an elegant formalism, but to solve a series of problems whose results need to be checked" (DeWitt 1957, p. 150).

There exists a large literature on the quadrupole formula from the period after the first announcement of orbital decay data from the binary pulsar. It will have to suffice here to discuss two of the most interesting new derivations of the quadrupole formula, those by Damour and Anderson, and then the two main opponents of the quadrupole formula during this later period, Rosenblum and Cooperstock. Damour's detailed analysis of the problem of motion, intended to compare directly with the experimental results from PSR 1913 + 16, is the closest thing to a solution to the quadrupole formula dispute, in the sense of its wholly or partly satisfying as many people as possible. Anderson's approach is regarded by a number of authorities as the most accessible and direct derivation of the quadrupole formula, because the validity of the formula was, to some extent, incidental to Damour's approach.

Damour, a product of the rather formal and mathematical French educational system, did his graduate work in Paris on classical renormalization theory in the context of a tensorial (gravitation-like) field. Before college, he had introduced himself to the problem of motion in general relativity by studying the EIH method. This early interest encouraged him to work on this type of problem as a student. After completing his thesis he was awarded a fellowship to go to Princeton and arrived there just before the announcement of the discovery of PSR 1913 + 16. With Remo Ruffini he quickly produced a paper on the new find, which was pessimistic that it would ever have implications for radiation effects. After that he did no further work on the radiation reaction problem until 1978, when he attended the Munich Texas symposium. The announcement at that meeting of data on orbital decay and the lively discussion on the state of play on the theoretical side encouraged him to address the problem. He joined a friend, Nathalie Deruelle; her advisor, Lluis Bel; and other collaborators in working on the foundations of a new fast-motion approach to the radiation damping problem (Bel, Damour, Deruelle, Ibanez, and Martin, 1981). This initial work restricted itself to the (unbound) scattering case for simplicity, as had Kovács and Thorne and Rosenblum, since the extra iteration (identified by Walker and Will [1980] as crucial) required by the bound orbits problem was very difficult in the fast-motion case. He began to develop his own approach to this problem, and throughout the 1980s he extended this work, always relating his efforts closely to the specific system presented by the binary pulsar. Working on his own or with Deruelle, and later with a student Luc Blanchet, he achieved a remarkable level of agreement between his problem of motion calculations and the observations of Taylor and collaborators on PSR 1913 + 16.

Damour, even in a field with a strong interest in its own history, had an extensive knowledge of the literature going back to Poincaré and, indeed, Laplace. In his 1982 paper presented at the Les Houches meeting on gravitational radiation, he situated himself in the fast-motion tradition in the problem of motion, or the post-Minkowski approximation (PMA), as he preferred to call it. It should be noted that by this time, however, any strict boundary which may have existed between slow- and fast-motion approximations was becoming somewhat blurred. Burke and Thorne's work had made it absolutely clear that the post-Newtonian approximations (PNA) were quite inappropriate in the far zone of the field, where a linear type of approximation was needed to correctly express the proper boundary conditions and matching techniques then used to apply them to the PNA solutions to the motion of the source in the near zone. Similarly, in order to avoid a heavy calculational burden, Damour truncated his expansions for the source motions in the near zone, restricting himself to slowly moving objects, such as the binary pulsar system itself. Therefore he, too, was left with a PNA-type of expansion for the equations of motion. Again, different types of expansion were more appropriate to different regions in the problem's geometry, with matching techniques typically employed to reconcile them.

One important similarity between Damour's method and EIH was the use of point sources. EIH employed a surface integral around the singularities to "cloak" them from view, so that only their field effect beyond the surface in question played a role in the problem, and the hidden objects could be presumed to be any body that would fit inside the surface and produce the same field. Damour preserved the cloaking effect, in order that his bodies could be presumed to be compact objects like the neutron star(s) in PSR 1913 + 16, but rejected EIH's surface integral method as too involved calculationally, substituting a volume integral instead. The use of point sources left him with the familiar problem (from electromagnetic field theory) of infinite integrals, which he solved by introducing a mid-century renormalization technique from electrodynamic theory, due to Marcel Riesz. This technique had been introduced into general relativity by Havas, who referred to the use of *Riesz potentials*. Damour preferred to use the term *analytic continuation*.

In certain overall respects, Damour's approach in his 1982 paper (along with much of the modern work on radiation) can also be compared to Fock's in the use of harmonic gauge conditions, imposition of the "no-incoming" boundary conditions, and matching techniques. But Damour's wide-ranging knowledge of the literature enabled him to sublimate various influences into an overall method that was distinctly his own. Besides EIH, other early contributors whose work he made use of were Asher Peres and Moshe Carmeli, another Israeli physicist who had done work on the fast-motion approach (pers. comm.). Another aspect of his assessment of the preexisting literature was his critical approach, which led him to make significant changes even where he was most inclined to imitate previous efforts, as with EIH.

A quite different approach to EIH and the previous literature is found in the late 1980s work of Jim Anderson. A student of Bergmann's, Anderson did not work on the radiation reaction problem until late in his career, when he entered the field during the quadrupole formula controversy of the late 1970s. Like Damour, Anderson was a strong critic of previous attempts to derive the quadrupole formula, describing some of them as "proof by naming," since they made use of an energy-balance argument relating the flux of energy in the wave zone to the loss of energy by the source system without establishing, in Anderson's view, that these quantities were really related. Nevertheless, Anderson did not reject energy-balance arguments out of hand and made use of them in his 1980 paper on the quadrupole formula. Possibly his most interesting paper on the subject, however, is his 1987 paper, in which he revived the EIH method, which in his view had been badly neglected by the field, and allied it to Burke's matched asymptotic expansion method and other applied mathematics techniques, such as multiple time-scale expansions, in order to deal with the problem of handling two quite different types of expansions simultaneously.

Anderson viewed the EIH paper as "arguably one of Einstein's greatest contributions to physics" (Anderson 1995). Largely because of the highly involved calculations that the method required, the EIH surface integrals were not employed by anyone other than Infeld himself, whose distrust of retarded potentials led him to reject the radiation terms in the expansion. Furthermore, in Anderson's view, the slow-motion approximation was inherently incapable of dealing with radiation anyway, and it was not until the work of Burke on matched asymptotic expansions that this failing was overcome. The virtue Anderson saw in the EIH approach was that by the use of surface integrals around point sources, it avoided the need for infinite-mass renormalization techniques (such as analytic continuation), which were required in field-theory problems to remove the infinite self-energy of a particle with no physical extension sitting in its own field. Anderson also addressed Infeld's criticism of the arbitrariness of using retarded potentials and showed that if "the energy of the initial field is finite, then in the asymptotic future the field is purely outgoing" (Anderson 1992, p. 466).

Anderson's "archaeological" use of the field's history, in which he constructed his new solution from preexisting elements in the literature stands in contrast with the largely rhetorical use of history in many of the review papers. Nevertheless, like many of the other reviewers of the literature in the '70s and '80s, he was quite critical of other procedures that purported to derive the quadrupole formula. He and Damour were highly critical of each other's calculations, for instance! Each was inclined to regard his own contribution as the only correct derivation of equations of motion for radiating systems. In this respect Anderson and Damour can be grouped with Ehlers, Havas, and Rosenblum as among those who insisted on the primacy of *method*: regarding the problem as one of finding the one best method that overcomes the important difficulties in principle. Thorne (1989), Cooperstock and Hobill (1982), and Walker and Will (1980) took a more relaxed view and regarded more than one existing calculation as containing positive features. Walker and Will argued that the quadrupole formula could be said to be reproducible when specific criteria are met and listed a sequence of existing calculations that met those requirements (as well as others that did not).

The radiation reaction work in the 1970s and '80s differed in one essential respect from that of previous decades. In most cases, the results of the various papers published agreed with each other and with the quadrupole formula. One does not need to look far to find a possible reason. By this time, both theoretical and experimental opinion had largely arrived at the conclusion that the quadrupole formula was the correct result to leading order. The corollary to this was not only the rejection of all other results, but the temptation to reject or view with deep suspicion any method or calculation that had led to conflicting results. In this period only two active researchers persisted in upholding contradictory results against the quadrupole formula. One was Arnold Rosenblum, whose fast-motion scattering calculation (disagreeing with that of Thorne and Kovács) gave a result, as quoted by Ehlers at the time of the ninth Texas symposium, that predicted energy radiated *in excess* of that predicted by the quadrupole formula by a factor of about 2.3. The other was Fred Cooperstock, whose rigid-strut model also predicted emission in excess of the value generated by the quadrupole formula. This represented a reversal of the position of the previous generation of skeptics, such as Havas, Rosenblum's mentor, who generally tended to suspect that the quadrupole formula *overestimated* energy loss by isolated systems.

Despite their fairly isolated position, both Rosenblum and Cooperstock proved to be vigorous advocates of their viewpoints. Neither shrank from public debate, and the exchanges grew quite vitriolic by the early 1980s. Both were forceful personalities with the rare ability to carry on a controversy when in the minority position. Both seem to have been able to leaven their debates with humor, which no doubt helped their cause. Cooperstock's students at the University of Victoria in Canada have even affectionately devoted a Web page to his humor, called *The Fred Cooperstock Experience*.

Nevertheless, if nothing else, weight of numbers began to tell. From the mid-eighties on, opinion had crystallized against the skeptics, so that there was little further debate already when Rosenblum died unexpectedly in 1991 at the early age of 47. This was a blow to the fast-motion program initiated in the fifties by Havas and Goldberg (and always advocated by Havas), for Rosenblum was its last very active exponent. As had been demonstrated several previous times in the history of the radiation reaction problem, the tenuous position of a minority research program in physics is prone to reversals brought on by historical accident in the form of personal crises or death of important figures in the minority camp. (Another tragically early death was that of Bill Burke, who died in a car accident in 1996 in California.)

Cooperstock largely gave up the unequal struggle, discouraged by the increasing intricacies to which he was led in attempting to refine his idealized model in the face of a large battery of critics. The inherent weakness of seeking to overcome one important problem of principle by introducing ad hoc initial conditions was revealed by the exponential increase in other problems of principle brought on as the model was developed. Once again, a minority research program encountered difficulties in keeping pace with its rivals for one reason or another.

Subsequently, Cooperstock completely abandoned his position of the early '80s, in preference for a return to the old skeptical stance: that the quadrupole formula was wrong because gravitational waves cannot carry energy at all, and therefore radiation damping does not exist for isolated systems. His position was based on his new energy localization hypothesis, which stated that to describe the local field energy in the absence of matter, one should choose only those coordinate systems which eliminated the pseudo-tensors and therefore appeared to leave no local field energy. In short, no gravitational field energy could be transmitted across a vacuum. However, Cooperstock still maintains that gravitational waves exist and are emitted by systems like the binary pulsar; they just do not cause dynamical decay in such systems. Furthermore, such waves could be detected, not by resonant bar detectors such as were used by Weber, but by the new interferometric detectors, which would observe the motion of test masses by the waves without, according to Cooperstock's analysis, receiving the input of any energy.[1]

Since Cooperstock's new hypothesis is a return, in some general sense, to the skeptical view of the mid-fifties, it is perhaps not surprising to find him addressing some of the arguments that were made during that era. In order to rebut the Bondi-Feynman thought experiment of 1957, he presented a new analysis of a simple gravitational wave absorber designed to show that it does not receive energy from the wave, despite the relative motion of its components (Cooperstock 1992). In a more recent paper, he attempted to show that a gravitational geon cannot exist in general relativity, since the ability of such a body to hold itself together depends on the gravitational waves that compose it having mass and therefore energy (Cooperstock, Faraoni, and Perry 1995). What is interesting about these papers is that they cast Cooperstock in the role of a historical revisionist, in the sense that he seeks an alternative reading of the older literature on the subject by exposing long-accepted results as faulty. I do not, incidentally, intend any pejorative connotation to the use of the term *revisionist*.

Given the historiographic conservativism which, it has already been observed, appears to play such a key role in the progressive view of science, it is perhaps not surprising that Cooperstock's efforts to rebut arguments from a previous era made in favor of gravitational waves carrying energy has met with little or no response to date from other relativists. Revisionism, generally resisted by any social grouping when applied to its own historical self-portrait, is perhaps especially strongly resisted by scientists. Cooperstock's radical rereading of history is nevertheless consistent with his earlier historical perspective, as expounded in Cooperstock and Hobill 1982. In that paper, noting the paucity of experimental support for general relativity theory, he stated that "since the true goal of physical theory is the description of the real world, it is thus particularly appropriate, with regard to gravitational theory, to nurture a spirit of skepticism. Surely this is a healthy ingredient for the growth of any science" (p. 362). He added that this spirit was "exemplified by Rosen." He went on to warn against the twin perils of the "optimist-visionaries" and the "mathematicians." The former should note that

> there is much more to general relativity than there is to be found in Maxwell theory and while optimism is admirable, it must be realistically tempered. On the other hand, the mathematicians forget that physics is not mathematics and that rigor is not an end in itself. Real progress in physics comes from that subtle interplay between experimental data, intuition, and the introduction of generalizing concepts

and principles. It is probably the dearth of experimental data which distorts the normal flow of progress and gives this discipline [GR] a flavor all of its own. (Cooperstock and Hobill 1982, pp. 362–363)

I earlier labeled the viewpoint expressed in these words as "anarchist," not in the political sense, but because it appears to reflect somewhat the views of Paul Feyerabend in *Against Method* (1988). Progress in physics, according to Cooperstock and Hobill, is the result of a "subtle interplay" between several factors, and one should resist the impulse to impose rigid programmatic schemes on its pursuit. In this, their view is a little closer than most physicists usually go towards Feyerabend's idea of an "anarchist theory of knowledge," in which "anything goes" if it works, just as Cooperstock and Hobill emphasized the practical necessity for skepticism if the "true goal of physical theory" is to be achieved. Cooperstock's efforts to reopen the old debate about energy localization, which flourished particularly strongly in the first few years of general relativity and which involved the first serious questioning of ideas like the Bondi-Feynman thought experiment, can certainly be seen as interesting from the historian's point of view. But Cooperstock's ideas received almost no attention from physicists. Only someone with a great deal of "social capital" as the sociologists call it (see Pinch 1977 and Bourdieu 1975 for discussions) could have revived this debate, and Cooperstock's capital had already been depleted by the earlier quadrupole formula controversy. Similarly, Joe Weber's efforts to claim a successful detection of gravitational waves from the 1987A supernova, which if true would have been a Nobel Prize–winning discovery, attracted surprisingly little controversy, because Weber's capital had declined so much that it was difficult for him to obtain a hearing (see Collins 2004 for a discussion).

By the late 1980s, the sudden lack of interest in what had once been important topics of debate does not reflect any decrease in interest in gravitational waves as a topic of research; rather, the reverse. Certainly, the great majority of relativists who remained interested in gravitational waves had plenty to occupy them in the late 1980s and early 1990s. During this period, the study of gravitational waves became a sizeable, important field for the first time in its history. The seed first planted by Weber around 1960, and whose erratic growth since would have given little grounds for optimism to the casual observer, matured and blossomed at last. In the United States, experimental groups at Caltech (led by Ron Drever, and later by Rochus Vogt) and MIT (led by Rai Weiss), with vigorous support from the theorist Kip Thorne, secured an unprecedented level of funding from the NSF for a large detector program, LIGO. Similar projects followed in Europe (the French-Italian VIRGO, the German-British GEO 600), Australia, and Japan, all in various stages of conceptualization or development. Several are now operating. Besides launching the study of gravitational waves into what can only be called "big science," these detector programs offered a new role to gravitational wave theorists. In a marked change from the early days of gravitational wave detection, these new detectors expect to make use of detailed theoretical predictions of signals from inspiraling binaries in order to filter the signal from the relatively strong detector noise.

One of the consequences of this need is a strong emphasis on the development of numerical techniques to allow the exact solution of Einstein's equations on supercomputers for the case of a binary black-hole system. One of those involved in this endeavor is Jeffrey Winicour, who according to Ehlers produced "the first rigorous version of a far-field quadrupole radiation law" (Ehlers 1987). Winicour, another graduate of Bergmann's Syracuse school (and who in addition had experience working for the air force with Goldberg), was motivated to work on this problem by Ehlers and succeeded in showing that the quadrupole formula (expressed in terms of explicitly Newtonian quantities) would be the Newtonian limit (letting the speed of light $c \to \infty$) of the Bondi news function in a rigorously defined, but physically appropriate, set of circumstances. Winicour feels that no one ever did produce a completely satisfactory answer to the problem of the quadrupole formula but thinks that the issue may in any case be superseded by the emergence of numerical relativity. As numerical relativity introduces a new way of looking at the field, and a new language in which to talk about it, he expects that the quadrupole formula problem will be forgotten anyway, as it may not be a problem one can formulate in a fully general relativistic manner (interview). This view of a fully realized general relativity theory finally emerging, freed of the encumbrance of Newtonian concepts and field-theory analogies, is intimately connected with the success or failure of the whole project of numerical relativity.

Begun in the 1970s by pioneers such as Larry Smarr (who went on to be the founding director of the U.S. National Center for Supercomputing

Applications at Urbana-Champaign in Illinois), numerical relativity has seen in the rapid advance of electronic computation in recent decades the potential that exact solutions of Einstein's equations can be achieved by exhaustive numerical methods even for cases such as binary black holes. The success of this endeavor may indeed play a critical role in the analysis of data from detectors such as LIGO. However, the field of numerical relativity has proven enormously difficult. Partly the reasons are sociological, as the large research groups required must nevertheless closely integrate their efforts in order to share code. This is a significant cultural shift required of a field previously consisting of very independent-minded researchers. But severe technical difficulties stand in the way also. One example of this takes us back to the perennial problem of coordinate systems. Most schemes for solving the binary black-hole problem numerically involve coordinate systems realized within the computer program as computational grids. This means that some of the information contained in the numerical calculations describes the actual physics about the system, while some of it is merely coordinate (or gauge) information about the structure of the particular grid. While the mathematics constrains the physical information to move across the grid at the speed of light, there is no such restriction on the gauge information. This is one reason why numerical relativity codes can suffer from major instabilities. In analytical calculations, done with paper and pencil, the bits of unphysical gauge information precisely cancel each other out, but in a numerical calculation, there will always be some error in the numbers so that the sums do not quite add up to zero. In this case the errors are propagated all over the grid at faster than light speeds. In the case of black holes this means that although the physical information present in the problem cannot escape from inside the black hole, the numerical errors generated inside can! This escaped error quickly builds up and the code crashes. To paraphrase Eddington, it seems that numerical error travels with the speed of thought, just as spurious gravitational waves do, and a supercomputer's speed of thought is blazingly fast. It is for reasons such as these that the leading numerical relativists did not look very happy, to my eyes, when called upon by Thorne to wager on the time frame in which they would complete their analysis of gravitational wave signals from binary black holes (this took place at the final meeting of the Binary Black Hole Grand Challenge Alliance in Austin, Texas in 1995; For more on numerical work related to gravitational waves, see Kennefick 2000).

We see from this how much the fate of the theorists is bound up with the success or failure of the new generation of gravitational wave detectors. It is hoped by the proponents of these experiments that they will be the first instruments of a new field of gravitational wave astronomy. One already hears the phrase *electromagnetic astronomers* used by such proponents to describe all previously existing astronomers. Whereas for thirty years, general relativity has flourished due to its increased relevance to other fields of physics and astronomy, some of its practitioners can now see it emerging as a strong field of physics in its own right, complete with an observation program that will provide direct insight into physical systems, such as black holes, which are exclusively the preserve of general relativity theory. In this hope one senses the view that to retain its independence as a separate body of theory, general relativity must progress or be eclipsed by more dynamic theoretical disciplines.

12

Keeping Up with the Speed of Thought

After all this talk of the contrast between the "rigorous" approach of "mathematicians," and the "intuitive" approach of "physicists," some comment on the meaning of these terms is in order. The word *mathematician* is used in a casual way in general relativity to describe both a professional affiliation and a style of doing physics. Some relativists really are mathematicians in the professional sense, by training or inclination, and general relativity is thought to attract more mathematicians than the other branches of theoretical physics. Indeed, one still sometimes finds, at least in Europe, relativity groups in the mathematics rather than the physics departments of universities. Nevertheless, most relativists are physicists by training, a few even coming from an astronomy or engineering background. Despite this, some of them are considered rather mathematical in their approach, so care has to be taken to understand what is meant by the use of this word in general relativity.

In an interesting article on standards of rigor in mathematics, Jaffe and Quinn quote Landau and Lifshitz on this topic, "No attempt has been made at mathematical rigor in the treatment, since this is anyhow illusory in theoretical physics" (Jaffe and Quinn 1993, p. 5). Jaffe and Quinn suggest that mathematics can play the role of experiment in theoretical physics in areas which are remote from actual experiment (citing string theory as an important example), and protest the attitude, exemplified in our account by Feynman, that accords little respect to the cleaning-up activities of the mathematicians, thus tending to discourage such valuable work. They note, "A relevant observation is that most theoretical physicists are quite respectful of their experimental counterparts," a statement many experimental physicists, including at least one vocal gravitational wave experimenter, would take issue with but which is probably true nonetheless. Jaffe and Quinn continue:

> Relations between physicists and mathematicians would be considerably easier if physicists would recognize mathematicians as "intellectual experimentalists" rather than think of them disdainfully as uselessly compulsive theorists. (p. 5)

In addition they wonder if a "theoretical" style of mathematics is possible, which would be a more intuitive and speculative, less rigorous, style of mathematics, akin to theoretical physics.

To describe a relativist as a physicist may seem straightforward, but again there are subtleties to the usage that are not easy to articulate, even for insiders. Broadly speaking, the physicists may be more inclined to relate general relativity theory to issues in other branches of theoretical physics and to carry on their work in the style of theoretical physics as it is done elsewhere. The mathematicians are more concerned with the development of general relativity according to its own historical logic, its own sense of where the interesting problems lie, and practice a style that reflects that of applied mathematics. Although there is no universally consistent description of what is meant by each of these styles, nevertheless, to the physics style we can attach the word *intuition*, and to the mathematical style, the word *rigor*. Loosely defined, by *rigor* we refer to the issue of standards of proof, and by *intuition*, we refer to an inner, possibly unconscious, instinct for finding correct answers to problems. This is what physicists often refer to as the *heuristic* method, a word that is related to the Greek word *eureka*, meaning "I found it." It is well known that the method of finding a mathematical proof can be quite mysterious and very different from the formal statement of that proof. In mathematics it is generally impermissible even to state a result without the proof attached and, in contrast to other areas of physics, in relativity quite a lot of important work involves the statement and proof of theorems. In physics, on the other hand, great credit can often be gained for stating a result that is later proved or experimentally confirmed by someone else.

One nice illustration of the difference between a mathematician's standard of proof and that of a physicist is given in Silvan S. Schweber's book on the history of QED, which provides a classic description of the differing styles of some famous physicists (1994, p. 464). (For a discussion of Einstein's style, which Michel Janssen has described as "opportunistic," see Janssen 2006 and Kennefick 2005.) In it he described Feynman's "proof" of Fermat's last theorem, until very recently the most famous unproven conjecture in the history of mathematics. Fermat's last theorem states that the equation $x^n + y^n = z^n$ cannot be satisfied for any integer values of x, y, and z, if $n > 2$. Feynman's proof was accomplished by showing that the probability of finding a counterexample (i.e., finding integers that *do* satisfy the equation for $n > 2$) is vanishingly small for values of n greater than the highest value of n for which the conjecture was known to be true (about $n = 100$; the difficulty in proving the theorem lies in doing so for all possible n). Feynman developed an integral that expressed the probability of finding a counterexample for a given n and concluded, since that probability was so low (less than 1 in 10^{200}) that "for my money Fermat's theorem is true." The mathematician's proof of this conjecture, when it finally came many years later, was notoriously complex (150 pages long). As an instance of why Feynman's approach may not be generally applicable to pure mathematics, Schweber noted that a mathematician had shown Feynman that his method could lead to wildly incorrect claims in some similar cases.

What is the role of intuition in physics? Arguments based explicitly on an appeal to intuition rarely make their appearance in published physics papers. If physicists do regularly make use of it in their work, then they are likely to censor its traces from their official accounts of how they make their discoveries, though less completely than is common for mathematicians. One might conclude that such inspiration or insight is distrusted by a community that regards itself as thoroughly part of the rational enlightenment tradition. The oral folklore of the community, however, paints a rather different picture. Stories of inspiration and intuitive leaps of reasoning abound in science in general (Archimedes in his bath, Kekulé and the benzene ring) and in physics in particular (Dirac's realization of the analogy between Poisson brackets and Heisenberg's commutators, which came to him "out of the blue"; Einstein's epiphany with the equivalence principle, "the happiest thought of my life"; both quoted in Chandrasekhar [1987, p. 20], who also relates interesting examples concerning Fermi and Heisenberg). Indeed, in personal interaction with other physicists, a physicist may reverse the procedure of a published paper and disguise rational deduction as inspirational insight. An example of this is given in the forward to Feynman's *Lectures on Gravitation* (1995) by John Preskill and Kip Thorne:

> Sometime in early 1963, Fred Hoyle gave a ... seminar [at Caltech] on the superstar model for strong radio sources. During the question period Richard Feynman objected that general relativistic effects would make all superstars unstable [to gravitational collapse]—at least if they were non-rotating. (Preskill and Thorne 1996)[1]

The substance of Feynman's objection was subsequently verified by several researchers. Preskill and Thorne continue:

> To Hoyle and Fowler [coauthor of the superstar model with Hoyle], Feynman's remark was a "bolt out of the blue," completely unanticipated and with no apparent basis except Feynman's amazing physical intuition. Fowler was so impressed, that he described the seminar and Feynman's insight to many colleagues around the world, adding one more (true) tale to the Feynman legend.
>
> Actually, Feynman's intuition did not come effortlessly. Here, as elsewhere, it was based in large measure on detailed calculations driven by Feynman's curiosity.

The evidence for these calculations is preserved in the notes for one of the lectures published in Feynman 1995 (the lecture was given shortly before the Hoyle seminar in 1963) and in Feynman's notes that are preserved in the archives at Caltech.

If the oral tradition of a subject is of value in illuminating the attitudes of scientists, it is particularly instructive to look at a special class of legends known as origin myths, stories that describe the genesis of a discipline. I regard textbook and memoir accounts as merely collected folktales which are passed on in a manner appropriate to a very literate community, but in an informal setting, as distinct from the professional literature of the journals. Physics is particularly rich in origin myths. A celebrated and classic example is the leaning-tower-of-Pisa experiment. It relates an incident that probably never happened but that defines something about the morals of the field. The ancestor figure, in this case Galileo, is depicted illustrating how one does science properly. The moral is not a surprising one: Galileo performs a critical experiment that at once destroys the old, opposing, fictitious theory and confirms his new one. It is not at all surprising that physicists should wish to portray their profession in this light, appealing to nature in an unambiguous way to confound their opponents. The mythic quality of the story is illustrated by the fact that it is far from clear that any single crucial experiment sufficed to destroy the reigning physics worldview that Galileo argued against.

In an equally famous origin myth, centered around another ancestor figure, we encounter a different moral. The story of Newton's apple actually derives from a real, contemporary anecdote, but it is very obviously a myth in the true sense of the word. In this story, the physicist is in repose, under a tree. By chance, an apple falls, and in a flash a great insight comes to him. This tale stands in remarkable contrast to that of tower of Pisa. The scientist sits under a tree, like a primitive sage. We might imagine it as the world tree, although it is not the world ash of northern European myth; instead, it reflects biblical symbolism. A vision of cosmic truth descends upon him from above (literally in many popular representations, where the apple is made to strike Newton on the head). The story quite clearly conveys the idea of inspiration striking the physicist, and this moment of intuitive insight (spoken of by other early modern theorists, such as Kepler[2]) is made to stand in, in mythic terms, for all the years of work that Newton was required to do to fully develop the concept of universal gravitation. Yet it also stresses the centrality of the moment of insight that gives direction to the whole process. The story, which is an exceptionally popular one among physicists in terms of retellings, quite consciously portrays the scientist as a seer, one who is inspired by divine powers. It seems fair to conclude that the inspirational mode of thought is one that physicists wish to claim for themselves by this origin myth, just as much as they would lay claim to the experimental investigation of nature in the Galileo story.[3]

A third important strand in the origin myths of the scientific revolution is the rationalist fable of Galileo confronting church superstition and dogma in his trial for heresy. Misleading as it is historically, the popular account of this episode claims the mantle of the rationalist tradition as part of the inheritance of physical science. The primacy of this same trinity, which is represented by the myths of the tower of Pisa, Newton's apple and Galileo's trial, is appealed to by Cooperstock and Hobill, as quoted earlier ("the subtle interplay between experimental data, physical intuition, and . . . generalizing concepts and principles").

This tripartite division of the art of science reflects a traditional division of the arts and sciences that can be observed as early as Francis Bacon, one of those early modern scholars who decried the neglect of useful knowledge in the philosophy of the ancients. In his classification of knowledge Bacon recognized three mental faculties, each of which governed a specific realm of the arts, namely memory for history, the imagination for poetry, and reason for philosophy, including natural philosophy. Of course Bacon understood that the faculties of the mind work in unison and play a role in all study; nevertheless in assigning to reason the governing role in what we would now call the physical sciences, he appealed to an old belief and at the same time anticipated an increasingly rigid demarcation. Many later scholars were influenced by Bacon's scheme, including the encylcopedists, who followed his general scheme in their organization of human knowledge. [Martinez 2001, pp. 16–21]

This trinitarian depiction of the art of science reflects ancient beliefs related to the threefold division of time. The deeper mythological associations of the origin myths of physics are perhaps indicated by the original names of the three Muses of Helicon, which have been translated as Meditation, Memory, and Song (Graves 1966). In other words, they represented the three indivisible strands of the poetic art. Memory is knowledge of the past, Meditation (or alternatively, Practice), the art of composition that lives in the present, and Song, the creative aspect of composition, suggests inspiration, in which the future form of the composition is "remembered." As studies have shown, oral epic poets remember the tale to be sung, but not verbatim, as we would understand it (Lord 2000). Instead, employing their repertoire of stock phrases and formulas, they improvise each performance as they sing it. Thus the overall shape of each rendition is remembered both in the past (the experience gained from previous performances) and in the future (because each performance is constructed uniquely as it is sung) with the employment of their learned technique. We can draw our own analogy with the work of a modern physicist, substituting experience for memory, mathematics and experiment for practice, and intuition for song. Experiment and other forms of technique (what the sociologist calls tacit knowledge) dwell in the present and inform all scientific work, even theoretical, since we must include here the battery of mathematical techniques employed by theorists in their calculations. A good deal of this analytic work is, however, rooted in the past, just as we have noted that formal mathematical analogies are of use to all physicists in helping them benefit from their experience with previous calculations or experiments. In the typical picture of the rational approach to physics, one builds towards the final answer in a series of steps, each sure and incontestable in itself. The more logically secure each step, the more rigorous is the calculation or proof. Intuition on the other hand, leaps into the future to recognize the form of the final answer before one has arrived there. As such, intuition is both highly suspect and indispensable to any mathematical or scientific discipline. At first glance it is much more dependent on experience than is the case with analytic technique, which pretends, formally at least, to be a logically complete demonstration, independent of analogy or other heuristic devices. Intuition is the guide that enables the researcher to pick a path through the thicket of possibilities to the solution, but it is not at all apparent that it will have selected the right solution. After all, whatever path was taken by the intuitive mind to that solution usually cannot be reconstructed. This is one good reason why intuitive insights are distrusted by scientists, because, as the philosophers say, the context of justification is different from the context of discovery. The arguments that make discoveries possible may not be well suited to subsequent justification. On the other hand, scientists are very familiar with the impressive role that intuition plays in their field. Thus a Feynman, already highly regarded for his abilities, may choose to add to his mystique by failing to reveal the source of his insight in terms of his equally amazing ability to calculate. It appears that it took Feynman some time before he realized that the superstars proposed by Hoyle and Fowler would actually be unstable to gravitational collapse. It just so happened that he had the advantage of having previously studied the problem in great detail.

Thus physicists lay claim to experimental practice, physical intuition or insight, and logical argument as three strands of their science in these origin myths. By their suppression of the intuitive strand in their professional discourse (i.e., in journal papers), they place themselves within the classical enlightenment tradition in which the three muses (later expanded to nine and given more strictly demarcated specialties, such as Urania, muse of astronomy, presumably the guiding muse of the scientists in our story) are symbolically subordinated to their mother, Mnemosyne, or Memory.

The alternative romantic view is expressed by Blake, in the prelude to *Milton*, "And the Daughters of Memory shall become the Daughters of Inspiration." Thus the upholders of mathematical rigor can be seen figuratively as the defenders of Heaven (and therefore both innocence and the tyranny of reason) in the face of the onslaught of Blake's forces of Hell (upholding experience and the power of the imagination).

The distinction between heaven and hell lies not in past versus future as such, but in an alteration in the relationship between the two in the shift from orality to literacy. Reliance on experience to inform intuition is a characteristically oral mode of thought. In literate, rationale modes of thought, memory is rendered more impersonal by being externalized and frozen in the form of writing. Rational thought tries to build a bridge from the known into the unknown, but in a characteristically different way from intuitive thought. Personal experience counts for nothing, because all arguments must be universal, they must apply to everyman (or everywoman), regardless of his (or her) personal experience. Thus the equation between innocence and reason. To become universal, a proof must divest itself of the individual experience, something intuition can never do. Intuition and experience can speak to another individual, but it cannot be read by the whole community. Although physics may seem like the most literate and rationale of subjects, in my experience it is in many ways a very oral culture.

With the help of this insight we can see how a calculation like Landau and Lifshitz's derivation of the quadrupole formula can provoke such widely divergent responses. Generated by a pair who regard rigor as "illusory" in physics, a proof that appeals to intuition and to experience strikes a powerful chord with those whose experience resonates with that of the authors. It leaves others cold and is viewed with deep suspicion by those of a more rigorous, mathematical bent. Jim Anderson, for instance, told me that Landau and Lifshitz's demonstration that binary stars emitted gravitational radiation according to the quadrupole formula was only credible "if you believe Landau was connected to God" (interview), while Bondi referred to it as "a little glib" and "full of holes" (interview). Advocates of Landau and Lifshitz's approach, such as Kip Thorne, have pointed out that the famously terse and concise Landau and Lifshitz style contributed to its mixed reception. Again, speaking in shorthand suggests that one is speaking only to those who are prepared to work their way through to their own insights.

If there are three poles to creative scientific work—analysis, technique, and insight—then it is reasonable to speculate that differences in emphasis as to which poles are important can characterize different disciplines within science. Different choices of which pole to weaken define the style of physics, since placing the emphasis on method emphasizes the rationalist project of science, whereas lending weight to the significance of the correct result inevitably leads to a reliance on one's intuition as to which result is actually the correct one. It is never possible to disentangle logic and intuition entirely in the process of science. For instance, Chandrasekhar insists on the balance between these qualities in the work of the greatest scientists. He quoted Fermi as saying that he would never believe a physical argument without a mathematical derivation nor would he believe the mathematics without a physical explanation (interview). Nevertheless, the style conflict that is frequently referred to by the protagonists in the quadrupole formula controversy seems to have its roots in a greater emphasis placed on one or another of these three ways of knowing.

These differences in style may be reflected in the different responses by different relativitsts to the basic analogy between gravitational and electromagnetic waves. Those who preferred a more physical approach wished to solve a difficult problem with a credible degree of certainty that they had gotten the right answer. In this endeavor, the analogy could be a useful guide to the right path, based on the experience gained from the electromagnetic case. If the calculation they were making appeared to violate expectations based on the analogy with the more familiar case, then it was suspicious. On the other hand, those who preferred a more rigorous approach wanted to prove that the answer was such-and-such, which naturally means eliminating all alternative possibilities. The contrast here is between the mathematician's proof of Fermat's last theorem, which proves beyond a doubt that no counterexamples are possible, and Feynman's physicist's "proof," which shows that counterexamples are just extremely unlikely. If you wish to hunt down and disprove the existence of all possible caveats, then the analogy may prove more useful when examined for possible breakdowns. By analyzing it for imperfections, one hopes to find the flaws in the argument and test it to destruction, by destroying it if necessary. In addition, many physicists are most attracted by the possibility of "new physics," by which is meant the possibility that the gravitational case is decisively different from the electromagnetic case. Both Einstein and

Bondi seem to have been attracted to the idea that gravitational waves do not exist because it would be a new and startling result for physics.

As we have seen, analogy is a powerful tool of the imagination, and from it the imagination can take off in many directions. Returning to the two physicists whose differing take on this analogy was illustrated in the opening chapter, we can see that were many similarities between Bondi and Wheeler. Both had made their name in other areas of physics, both came to relativity theory in search of something new, both were gifted with great physical intuition and mathematical skills, and both had extraordinary influence on younger physicists. Both men were noted for their daring hypotheses, and both were prepared to defy conventional wisdom. Their techniques were, in many respects, similar, yet how comically divergent were their remarks about the analogy with electromagnetism in 1957 at Chapel Hill, a possible indication of profound differences in outlook between the schools of which they were a part. My own impression is that the British school upon which Bondi had so much influence was more mathematical and rigorous in its approach than Wheeler's school, noted for its more astrophysical approach; theoretical astrophysics is probably the most heuristic and least rigorous of all fields of theoretical physics.

How is it that the controversy over the existence of gravitational waves and whether they are emitted by binary stars could go on so long, indeed for many decades (when the various stages of the dispute are added up)? Controversy was not confined only to the theoretical side of gravitational waves physics, either. The heated exchanges over the experimental detection of the waves burned even more fiercely, and the leading figure in that history, Joe Weber, went to his grave insisting that he had detected them, in defiance of every other expert opinion. Why are these controversies (and these controversialists) so tenacious?

In his study of the Weber controversy, Harry Collins was led to propose the existence of the experimenters' regress, which describes a situation that may develop when experiments that attempt to measure the same thing disagree. If the only way to determine which experiment is correct is to see which one got the right result, and the only way to tell which result is right is by performing the "right" experiment, then the experimenters are caught in a vicious circle. The problem is exacerbated, and the duration of the controversy greatly lengthened, by the fact that there is no known way to infallibly discriminate between the experiments by inspection. The reason is that only the experimenters themselves know what is happening in their apparatus, and much of their knowledge is not easily articulated. What is required to perform an experiment correctly is what sociologists call tacit knowledge, the sort of knowledge one does not think to mention. Riding a bike is the classic example of tacit knowledge, because the most important characteristic of this kind of knowledge is that it is learned by watching and doing. It is big mistake to try to learn to ride a bike from a manual or via instruction over the phone. Similarly, the way to learn to perform an experiment is to go and do it yourself, but preferably in the company of someone who is already an expert.

Since much of what experimenters know is tacit knowledge, in times of controversy it can be very difficult to determine who is doing the experiment correctly. Not only will the debate focus on details not normally given a thorough airing in public, but the true origins of the problem may concern some detail of the experimental method that the experimenters themselves are not aware is of any great importance. Collins (2001, 2004) tells the revealing tale of a Russian gravitational wave experimental group who had achieved unprecedented success in suspending crystals by wires with very low experimental noise. Since other groups were unable to replicate their results, there was some suspicion that the Russians were exaggerating their achievement. Finally a foreign group sent one member to Russia to work with the locals, and it was eventually figured out that the problem lay in the failure of the other groups to grease their wires. The Russians typically did this by rubbing the wire behind their ear before installing it and understandably had not mentioned this step in their published reports!

Because of the shear wealth of detail and the unconscious nature of some of the knowledge involved in doing an experiment, controversies may not be easily settled even by attempting to replicate the experiment. It is sometimes imagined that replicating experiments is one of the distinguishing features of science, yet in practice, true replication is very rare and probably always problematic. There is no great need to replicate an uncontroversial experiment (classroom replications of famous experiments do not count, since every contradictory result is simply put down to the incompetence of the student and thrown away). In cases of controversy, a replication of an experiment is open to all the same challenges as the original. Someone will say, yes you tried to replicate the experiment, but you did it wrong. The number of possible ways of doing the experiment wrong is, if not infinite, at least sufficiently great to exhaust the patience of physicists anxious to reconcile the progress of science with the desire to give everyone a fair hearing. At some point in long-lasting controversies there comes a moment when the community as a whole starts tuning people out. This moment can seem surprisingly sudden after a long and carefully argued debate. Because it is the result of a complex social interplay within a large community, the reasons behind it may even be quite mysterious, or even capricious. Not everyone is convinced by quite the same argument, but eventually enough people agree about which result is correct that the details no longer matter.

The best-known argument against the existence of the experimenters' regress is due to the physicist and philosopher Allan Franklin who, discussing Collin's history of the Weber controversy, argues that Weber's critics were justified in cutting off the debate as they did, ceasing to listen to Weber's arguments. He believes, quite rightly, that they acted reasonably in doing so, based on the wealth of arguments they had made against Weber's position (Franklin 1994). I would say, however, in agreement with Collins (1994), that Weber also behaved reasonably in continuing to argue his case. Since no one ever showed precisely what was wrong with Weber's experiment, it was certainly open to him to continue to interpret his results as being consistent with gravitational wave detection, and indeed, his stubborn refusal to recant his views was exactly the quality that had permitted him to succeed in his career up to that point. I note that the principal argument in favor of Weber accepting defeat, as he was urged to do by many colleagues, was essentially social in nature, that is, that he was harming his professional standing by his insistence on defying the overwhelming majority of his peers.

It is worthwhile to recall that there have certainly been cases in the history of science when reasonable arguments pointed in the direction of the wrong answer. In the debate over the Copernican model of the solar system, for instance, skeptics, such as Tycho Brahe, observed that the motion of the Earth ought to be detectable by measuring parallax shifts in the position of the fixed stars relative to each other. The answer that the stars were too far away for this shift to be observed, while correct, must have seemed to the skeptics like a classic ad hoc attempt to dodge the result of a *experimentum crucius*. The truth is that there are many lines of reasoning which are voiced in any given debate, and never is life so accommodating as to line up all of them up with each other. Since different people give different weight to the different strands, it is inevitable that perfectly reasonably people manage to reasonably disagree. Alas, they do not always agreeably disagree. I may add in passing that the philosophical debate over the nature of the experimenters' regress no longer seems quite so black or white as once it did, and as a practical matter, there is not much difference of opinion between Franklin and Collins in their views on the Weber controversy.

But is the case of theory different? Surely calculations are more transparent than experiments, without the quirks of a physical apparatus to be allowed for. In fact, theorists also have their tacit knowledge to acquire, and each of them performs calculations in their own way. Replication may be somewhat easier than is the case with experiments, one needs fewer screwdrivers and more pencils at any rate, but it is still problematic and prone to provoke arguments when the results do not agree. The number of steps in a serious physics calculation may not, thank God, be infinite, but there are more than enough of them to keep a discussion going for a very long time. Now that many calculations in subjects like relativity are conducted partly or wholly on computer, the analogy with experiment is even more starkly clear (Kennefick 2000). The computer code is a virtual apparatus whose inner workings are best known to the theorist who wrote the code, and not even completely to him or her. As we have seen, even in the purely pen-and-paper days, things of which the theorist was unaware were going on inside the calculations, as in the case of Peres's discovery of ingoing waves impinging on the source that he had not intended.

So in a case where the test of a correct calculation is that it gets the right answer, but yet the point of doing the calculation was to establish the identity of the right answer, the "theoreticians' regress" may be encountered. When the results of two or more calculations do not agree, one descends into a vicious spiral of point and counterpoint as each calculation or method is subjected to a seemingly endless critique. In the end, when enough people have made up their minds one way or the other, the debate gets tuned out and eventually peters out, leaving a few dissatisfied but freeing the majority to move onto the next problem.

But the role of intuition, and therefore the relevance of style should be obvious, since the easiest way to break out of the regress is to decide which of the possible answers is the correct one. The argument over which calculation is correct may be virtually endless, but once you have decided which answer is more likely to be correct, then the calculational method or methods which tend to recover that result can be trusted more than those which do not. Keep in mind that, as with experimenters, theorists need to check their calculations against known and accepted results in order to have faith that they are on the right track. Intuition, which, as I have described it above, is essentially the knack of guessing the right answer ahead of time, is thus likely to play a role whenever the theoreticians' regress raises its head. It is not suprising, in the case of gravitational waves, that the controversy was largely viewed as a struggle between intuitive physicists who accepted the validity of calculations that recovered the quadrupole formula versus rigorous mathematicians who were highly suspicious that this was a case where mere analogy was leading the calculations towards a plausible but erroneous answer. For example, to the typical "physicist," the fact that, from some point in the sixties on, many calculations of different types agreed upon the same result, was reassurance that said result was in truth the correct one. For the more mathematically minded, the agreement between many different results that were all conceptually dubious was merely evidence that the authors of those results knew what they were looking for.

The question of how and when theory ends is, just as with the question of when experiments end (Galison 1987), essentially a question of how the scientist knows when he or she has found the right answer. It is always possible to make a thorough check and find another calculational error or another bug in the code that might change the result. At some point, one must decide that the current answer is the right one. After all the checking of intermediate results and special cases against previous work, one comes to the point in research at the frontier when a new result must be presented. Every theorist we have discussed checked every aspect of their work very carefully, but eventually they all had to make a decision to stop checking. For a skeptic like Havas, the fact that he and two coauthors had at one time all agreed on the sign of their result but had made independent (and different) errors to reach that result must surely have been an instructive lesson in the perils of ending theoretical work too soon. Recall also Anderson's criticism of "proof by naming," discussed earlier. Certainly there were cases where the quadrupole formula was recovered essentially by accident, but nevertheless, the accumulation of results grew more impressive as time went on. This is not to say that the individuals concerned were that impressed by sheer numbers, but enough different paths leading to the same result can satisfy a sufficient number of different criteria to achieve a critical mass of convinced physicists. Once this stage is reached, the remaining unconvinced participants will no longer receive the same hearing they once did.

An event that helped to partially break the cycle in the case of the quadrupole formula was the advent of the binary pulsar data. Of course it is hardly surprising that an experimental or observational result could provide a way out of the theoreticians' regress, but interestingly this data gave rise, in the first instance, to more activity and more disagreement. The data lent critical support to the preferred "right result" given by the quadrupole formula even though it did not put an end to disagreements about the correctness of various methods, except in so far as it tended to rule out methods that did not derive the canonical result. This support was enough to gradually bring an end to the public side of the quadrupole formula controversy. Once a result is no longer in dispute, most observers to not care so much for the details of which calculation is most correct. Though there were private doubts among some theorists as to whether this was really the last word, and though it has taken some time for a confirming observation to come along, the old controversies never resurfaced once the debate was closed.

The role of social constraints within a community of theorists in the absence of experimental data is nicely illustrated by the course of the quadrupole formula problem. Keep in mind that general relativity's former characteristic of being a field with "too much theory in pursuit of too little experiment"(Bonnor 1963, p. 555),[4] once a pathology within physics, is increasingly becoming normative. String theory, for instance, flourishes as a far more mainstream and larger field than general relativity ever did and has even less contact with the world of experiment. The question of how one does physics in the absence of either mathematical or experimental rigor is a very live one for string theorists (as Jaffe and Quinn discuss in their 1993 article, mentioned above).

To begin with, in the quadrupole formula controversy, almost no constraints on acceptable answers (in the sense of *publishable* results) were in effect, even if they were inherently "unbelievable," in Peres's words. Peres faced strong social pressures encouraging him to accept his first, incorrect, result. The problem constituted his thesis project, and the desire to graduate was a strong motivation to finish the calculation. It was only after his successful doctoral defense that he discovered the flaw in his algorithm (Peres, pers. comm.). Similarly Hu, who had presented his result in public before he discovered an error that changed the sign of the answer prior to publication, and Havas, for whom the strange result was emblematic of the unsatisfactory state of the field, had their own reasons for publishing a result that ran counter to all expectations. Nevertheless, it is true that from the mid-sixties on, no further energy gain results were published. It seems clear that in the wake of the successful efforts to describe the asymptotic behavior of the radiation, the field was no longer as wide open to interpretation as it had been. The space for acceptable results had shrunk somewhat. The answer to the question, what amount of energy is emitted by a self-gravitating system in the form of gravitational waves? now had a "right" sign and a "wrong" sign. Hand-waving arguments could no longer be easily employed to dress up such a result with an air of plausibility. Earlier examples include those of Hu, relating his expanding orbit result to the Hubble cosmological expansion, and Peres, drawing on a "disanalogy" with electromagnetism and observing that the sign of the gravitational attraction is opposite to that of electromagnetism, that is, like charges attract instead of repelling (and thus the field-energy density in the binary is negative, and the sign of the change in energy is reversed from the electromagnetism case). Admittedly, Peres's argument that gravitational waves might carry negative energy was still being advanced by Rosen in 1964. In fact, Rosen's bimetric theory, in common with some other minority gravitational theories, has been shown to predict orbital expansion rather than decay as the result of gravitational wave emission (Weisberg and Taylor 1981). However, such a feature of a gravitational theory is now considered to be a definite argument against its validity.

Nevertheless, although the space for legitimate disagreement had narrowed somewhat by the early 1970s, there was still ample room left. In the '70s, Rosenblum and Cooperstock disputed the field with the supporters of the quadrupole formula result. Furthermore, the agnostic attitude typified by Havas gained new advocates, such as Ehlers. Only with the advent of the binary pulsar data could the acceptable field of results be narrowed down more or less completely to one option. However, this last chapter of the story hardly took place overnight. It was only gradually that the space for dissent was worn down, and this was not achieved without some damage to reputations in the final stages. In the early period, dissent was permissible, whereas in the 1980s, it was no longer the case. There seems to be some evidence that Cooperstock, although he is still professionally very active, lost a certain amount of standing on this topic as a result of his open defiance of the emerging orthodoxy. Whether this would have been true also of Rosenblum had he lived is hard to say, but it seems quite possible. Certainly, the practical effect of Cooperstock's loss of standing may be seen in his inability to provoke debate over his new challenge to the establishment, in which he revives the old argument about whether gravitational waves can carry energy and about the role of the pseudo-tensor. That avenues of expression still remain is evident by the publication of the initial paper on his energy-localization hypothesis in *Foundations of Physics*, a journal that encourages the publication of "speculation not tied to hard and demonstrable facts [but] suggestive of new basic approaches in physics" (editorial preface, *Foundations of Physics* **1**(1), 3), in which Joseph Weber has also published in recent years. Nevertheless, as discussed above in chapter 11, provoking a rejoinder to initiate debate often proves more difficult than simply gaining a platform.

Finally, even when the point was reached that room for disagreement on the leading-order *result* for gravitational back-reaction was eliminated, the same could hardly be said of the question of method. For those for whom this was always a central consideration (Havas, Ehlers, Damour, Anderson) there is still much to be critical of in all, or all but one, of the multitude of derivations of the quadrupole formula. Damour, indeed, resists the tendency to reduce the problem to one of verifying the quadrupole formula (interview), which distracts from the larger question of the correct approach to the problem of motion in binary systems. Ehlers and Havas, when interviewed, expressed themselves as still unhappy with some aspects of the state of the field. Ehlers felt that enough work had now been done at least to convince him of the approximate validity of the quadrupole formula. Havas still seemed to entertain certain reservations on that score. Damour and Anderson both continued to be critical of all solutions to the binary back-reaction problem except their own, not least each other's. However, in a context in which the final answer is the same, such disagreements in principle are insufficient to sustain a public controversy.

It seems that the aim of the all of the conferences, workshops, papers, reviews, appeals to experiment, and so on, is not to enforce or encourage *agreement* as such, but rather to eliminate or reduce the space for *disagreement.* There may be a distinction between what the community can agree to disagree about and what they must argue out. In the early period (1945–1965 or so), the overriding issue of principle, whether gravitational waves existed or were emitted by binary star systems, was something that had to be argued out. In the prewar period, although the issue was noted by Eddington, the subject had not matured to the point where it could support such a debate. After some point during the early 1960s, the debate ceased to be relevant because a sufficient consensus had formed against the skeptical position. Subsequent attempts to raise this issue by Havas, Rosen, or Cooperstock have received little attention. Similarly, the quadrupole formula controversy did not really emerge in its own right until after 1965. Up to that time the subject could not sustain such a debate, since the level of technical certainty or proficiency that could sustain a single canonical result had not really crystallized. But by the 1970s, the various post-Newtonian methods having at least agreed with other approaches on a consistent basis, there was sufficient ground for a debate over the status of the quadrupole formula. The advent of directly relevant experimental data, on top of the increasing astrophysical relevance of compact objects and gravitational waves, lent urgency to the matter. But only gradually was the room for dissent squeezed further. Eventually a critical mass of consensus, enough to close off further debate, formed in favor of the wide applicability of the quadrupole formula. As the field of gravitational waves moved into a new era, in which detailed calculations going beyond the quadrupole formula would be required for present (interpretation of the binary pulsar data) and future (gravitational wave detectors) experimental efforts, dissent was no longer viewed as healthy or desirable. Further disagreement would only retard the progress of a field which was showing signs that it was about to take a significant step towards the forefront of physics.

The controversial history of this field is interesting precisely because the persistent nature of the debate forced the participants to record their opinions, even if only as rhetorical weapons against each other. It is tempting to view the history of this problem as pathological, as Feynman was inclined to do, for instance, when he said that "the good men are occupied elsewhere" (see chapter 8). While this might be true, even the good men may follow a circuitous route, much longer than apparently necessary, to the solution of a problem. Feynman himself knew very well that scientists are very clever at covering up their tracks. As William Blake said in one of his *Proverbs of Hell* from *The Marriage of Heaven and Hell*, "Improvement makes strait roads, but the crooked roads without Improvement are roads of Genius." Many men of great talent, and a few geniuses, contributed to the development of gravitational wave theory. The fact that the history was controversial, as well as convoluted, gives us a rare opportunity to study the tracks of their genius undisturbed. That physicists themselves display so little interest in such crooked tracks is a feature of their "daring conservatism," already alluded to. They are always looking forwards rather than backwards. It seems to me, as I noted earlier, that this effacing of the past is part of the attraction that physics has for many very creative young scientists. Physics is always beginning anew for each generation, yet still advancing as it uses yesterday's results as mere tools, shorn of any unnecessary conceptual baggage that preoccupied the previous generation. Physicists, like everyone else, travel fast when they travel on straight roads.

This book's title refers to Eddington's famous phrase describing the illusory quality of coordinate, or gauge, waves, which can travel arbitrarily quickly, just as fast as you want them to. The speed of thought is indeed swift, proverbially so. In Snorri Sturluson's *Prose Edda*, one of the most famous stories is *Thor's Journey to Utgard*, in which the strong but not too bright Thunder God is tricked and humiliated by the illusions of the giant wizard Utgard-Loki. In a series of contests Thor and his companions are defeated by opponents who are later revealed to be manifestations of immaterial concepts. Thus Thor's servant Thjalfi, who is exceptionally fast, is defeated in a running race by the giant Hugi. But Hugi was not what he seemed, for the name means "thought" in old Norse (note also Odin's ravens, Huginn and Muninn, thought and memory, whose names again recall the division between different ways of learning about the world). Later Utgard-Loki confesses, "When Thjalfi ran the race with him called Hugi, that was my 'thought,' and it was not to be expected of Thjalfi that he should match swiftness with it" (Arthur G. Brodeur, trans., *Prose Edda* [New York: American Scandanavian Foundation, 1916]).

Eddington's skeptical remarks about gravitational waves that travel at the speed of thought capture well the flavor of much of the skepticism which marked this subject. It was a shrewd, intelligent, probing skepticism that illustrates much of what is best about the scientific method. But from the very beginning, this skepticism sometimes invited an impatient response.

Richard Feynman was one of those who thought that too much caution was inhibiting the pace of research in the field. At the Chapel Hill conference in 1957 he had warned, "Don't be so rigorous or you will not succeed," adding, "In this field since we are not pushed by experiments we must be pulled by imagination."

How does one account for the discrepancy between the extraordinary swiftness of thought displayed by physicists in this field in their individual work, skeptics and nonskeptics alike, and yet the extraordinarily slow pace of progress over the decades, which saw the validity of Einstein's work of 1918 still in debate in the 1980s? It seems that the speed of thought is, indeed, highly variable. On the one hand, individuals can shoot off at great speed in all directions, on the other hand, the great wave front of the community's knowledge can move with glacial slowness.

One has to keep in mind that it is the speed of the community's thought that matters. The speed of light can be defined in various ways to give various values (group velocities and phase velocities, and so on). This often perplexes those who have heard from Einstein that the speed of light is an absolute, or at least invariant quantity. What does it mean to say that light can travel faster than light speed? Is the speed of light really a limit? The argument goes that it is a limit on the speed at which information may travel. Information is the one thing that cannot travel faster than the speed of light, according to the theory of relativity. Parts of a light wave can do so, but any information carried by the wave will travel at the limited light speed. In science, it is the speed of the community's thought that matters (the "group velocity," one might say). However quick an individual is, if their idea is "too far ahead of its time," it will have little impact until the time is right. When the group or the community catches up, then the idea can be absorbed. Of course different communities have different speeds of thought, just as light travels much more slowly when passing through a pane of glass. It all depends on the medium. In the early days of a community, such as the one described in this story, progress may indeed appear slow, because much of the effort is going into building a coherent community, as well as constructing a language, a worldview, and other things that a scientific community needs. Only when a community has a shared experience and a collective memory (which includes its written memory, in the form of literature dedicated to that field), only when it is more than a collection of individual experiences and quirky intuitions,

Figure 12.1. Charles Misner, John Wheeler and Kip Thorne with their celebrated textbook, *Gravitation*. (Photograph and Copyright Anna Zytkow)

can it expect to make rapid progress with widespread agreement on what has been achieved. This, among many other things, means addressing the question of what standards of proof and rigor are appropriate to that field. While relativity has much higher standards of rigor than many other areas of theoretical physics, it has much lower standards than are prevalent in mathematics.

One of the features that marked the early days of the general relativity community was a certain tension between the "mathematicians" and the "physicists," between those who demanded a more mathematically "rigorous" approach to the subject and those who, like Feynman, wished to be "pulled by imagination." There is a phrase of William Blake's that can be taken as a motto for this field. It is one of his *Proverbs of Hell* and is quoted at the end of the bible of gravitation theory, the mammoth textbook written by Wheeler, Thorne, and Misner. It goes, "What is now proved was once only imagin'd." The speed of thought runs swift in the imagining and slow in the proving, but it gets to both in the end.

APPENDIX A

The Referee's Report

Included here is a transcript of H. P. Robertson's report to the *Physical Review* as the referee of the paper submitted by Einstein and Rosen in 1936 and discussed at length in chapter 5. Two copies of the report have been preserved. One is in the Albert Einstein Archives at the Hebrew University of Jerusalem (EA 19-090), the other, a carbon copy of the first, is at the Robertson papers in the California Institute of Technology Archive, Pasadena, California (box 7.12). The cover page that begins this transcript is only available at the Robertson papers, the rest of the pages (begining with "Comments of the Referee" below) are preserved in both archives.

Two comments are in order. First, Robertson is quite correct that much of the early part of the paper repeated results found in Eddington 1922. Einstein may, however, have been irked by this comment, since he was essentially also repeating his own discussion of 1918 (see chapter 4). Second, it is noteworthy how Robertson's intuition that plane waves are probably physically allowable and that the worries entertained by Einstein and Rosen are most likely "difficulties . . . involving space-time as a whole, rather than anything that may crop up in the analysis of a limited region of space-time" were very prescient, in view of the postwar settlement of this issue (see Bondi, Pirani, and Robinson 1959). Note also that the reference to a paper by Eddington, Baldwin, and Jeffery on the last page of the report is actually a reference to Baldwin and Jeffery 1926:

I [Cover Page]

If general thesis is correct, work is of first rate important. But I do not believe authors have established their case—see attached "Comments of Referee."

Typographical: see Comments (a), (b). Logical: see Comments (d), (e), (f).

Part I could be shortened without loss by refering to previous results—see Comment (c).

[comment e] If desired, "complete wave" solution could be taken over bodily from previous work but that would leave "incomplete wave" hanging.

Analysis is unexceptionable up to middle of p. 13, but I cannot agree with the crucial analysis on last half of p. 13, first half of p. 14, middle of p. 15; and without this the whole point of the paper is lost. See detailed "Comments of Referee" attached hereto.

Specifically, I maintain that the field defined by my eqs. (v), (vi), satisfying eq. (31)–(34) of the authors' paper, is a "Gegenbeispiel" which shows their argument to be fallacious. See Comment (e).

Referred back to authors for consideration of referee's criticisms.

I [Page 1]

Comments of Referee

Author: A. Einstein and N. Rosen.

Title: Do Gravitational Waves Exist?

(a) P. 2, eq. (3): $\overline{\gamma}_{\alpha\alpha,\mu\nu}$ should be replaced by $\delta_{\mu\nu}\overline{\gamma}_{\alpha\beta,\alpha\beta}$

(b) P. 2, eq. (6): Replace $+$ signs by $-$ signs, to accord with the $+$ sign in eq. (4).

(c) Pp. 3–5: That pure longitudinal and longitudinal-transverse waves are spurious has been shown by A. S. Eddington, *The Propagation of Gravitational Waves*, Proc. Roy. Soc. 102A, p. 268, 1923, to which no reference is made. Present work goes somewhat further than Eddington's in showing explicitly how to set up the infinitesimal transformation which removes them. It is to be noted that there exists a (rather trivial) class of spurious waves of type (c) for which the author's condition $\overline{\gamma}_{22} = \overline{\gamma}_{33}$ is not fulfilled—see (xii) below.

(d) Pp. 8–15. In Part II the authors seek a field

$$\text{(i)} \quad ds^2 = A(dx_4^2 - dx_1^2) - B dx_2^2 - C dx_3^2$$

which is to be the rigorous solution of the problem of plane waves, in which the coordinates x_μ are to correspond to the Cartesian coordinates employed in the derivation of the results of Part I. Now we should expect this to imply that (i) should contain, as that special field in which the tensor $T_{\mu\nu}$ vanishes everywhere, the solution

$$\text{(ii)} \quad ds^2 = dx_4^2 - dx_1^2 - dx_2^2 - dx_3^2$$

But the normalization which they adopt (eq. (23a) and that below eq. (30b)) preclude this, for in each case they have in essence used $\gamma = \log BC$ as one of the space-time *variables*. This use of a variable condition in the field, due to the presence of matter, in mapping the field may of course cause some difficulty in making the transition to weak or vanishing fields—although I do not consider this as particularly serious.

I note in passing that, in the case of the "complete wave" solution defined by (23a), (31)–(34), the only distinct empty-space solutions contained therein are

$$\text{(iii)} \quad ds^2 = dx_4^2 - dx_1^2 - x_1^2 dx_2^2 - dx_3^2$$

and the monstrosity

$$\text{(iv)} \quad ds^2 = \frac{dx_4^2 - dx_1^2}{\sqrt{x_4^2 - x_1^2}} - \frac{x_1^2}{x_4 + \sqrt{x_4^2 - x_1^2}} dx_2^2 - (x_4 + \sqrt{x_4^2 - x_1^2}) dx_3^2,$$

of which no more will be said. The first of these would seem to indicate that x_2 should be interpreted as an azimuth instead of a Cartesian coordinate—are we perhaps here dealing with cylindrical, instead of plane, waves? In any case it is clear that the situation cannot be saved by any of the allowable transformations (26), which preserve the "plane" character of the wave—it requires transformations involving at least one of the remaining variables x_2, x_3.

I have not thought it worth while to carry out the corresponding investigation for the "incomplete wave" solution (38)–(41), as I have treated it in (f) below under a seemingly more appropriate normalization.

(e) Pp. 12–14. The "complete wave." This solution is *mathematically* equivalent to the Weyl-Levi Civita axially symmetric static field (cf. Weyl, *Raum-Zeit-Materie*, ed. 5, p. 266 for solution and references—note next to last line contains misprint $h = e^\gamma$ for $h = e^\gamma/f$, and first term of right-hand side of last equation should read $2r\frac{\delta\Phi}{\delta r}\frac{\delta\Phi}{\delta z}dz$). In fact, the authors' "complete wave" may be obtained therefrom by the substitution

$$r = \bar{x}_1, z = i\bar{x}_0, \theta = \bar{x}_2, t = i\bar{x}_3;$$

$$\Phi = -\frac{1}{4}\beta \quad \frac{1}{2}\log\bar{x}_1,$$

$$\text{Weyl's} \quad \gamma = -\frac{1}{4}\beta + \frac{1}{2}\log\bar{x}_1 + \frac{1}{2}\delta.$$

The potential equation satisfied by Φ becomes the eq. (31), and Weyl's equation for $d\gamma$ yields (32), (33); eq. (34) is a consequence of (31)–(33).

It is not clear to the referee why the authors, in order to obtain a periodic solution, demand that the canonical solution be subjected to a transformation involving the periodic functions f, g; may not the canonical solution itself suffice for the discussion? But instead of attempting to explain the apparently paradoxical behavior of the field at events at which the Jacobian $4f'g'$ of the transformation vanishes, we merely point out that the complete wave solution does contain solutions which are adequately represented by the approximate fields, discussed in Part I, in regions where the field is "weak"—and would further maintain that this is indeed the general situation. The example chosen may be objected to on other grounds—namely, because of its behaviour at ∞ or because, as pointed out in (d) above, for vanishing $T_{\mu\nu}$ it approaches the empty-space form (iii) instead of (ii),—but it is to be noted that these are not the questions which enter in the authors' arguments (last half of p. 13 and first half of p. 14) (At least the first of these objections cannot be raised against the form of "incomplete wave" given in (f) below.)

Eq. (31) is that of cylindrical waves in ordinary space; we consider the comparatively simple solution

$$(v) \qquad \beta = 2\log x + \lambda J_0(\omega x)\cos\omega t + \text{const.},$$

where λ is a parameter, J_0 the Bessel function of order 0, and we here write $\bar{x}_1 = x$, $\bar{x}_4 = t$. The term $2\log x$ has been added on to assure that space-time is empty on allowing $\lambda \to 0$. The equations (32), (33) for δ—whose condition of integratbility is the eq. (31) itself—may then be integrated to give

$$(vi) \qquad \delta = \frac{1}{2}\lambda J_0(\omega x)\cos\omega t$$
$$+ \frac{\lambda^2}{16}(F(x) + \omega x J_0'(\omega x)J_0'(\omega x)\cos 2\omega t),$$

where

$$F(x) = \int^{\omega x} z[(J_0'(z))^2 + (J_0(z))^2]dz.$$

Since we are here interested in those portions of the field which approximate plane waves, we must take $\omega x \gg 1$; the Bessel function has here the asymptotic expansion

$$J_0(z) = \sqrt{\frac{2}{\pi z}}\cos(z - \frac{1}{4}\pi) + O|z|^{-3/2}.$$

Hence in such regions (v), (vi) become

$$\beta \sim \text{const.} + 2\log x + \lambda\sqrt{\frac{2}{\pi\omega x}}\cos(\omega x - \frac{1}{4}\pi)\cos\omega t,$$

(vii) $\delta \sim \text{const.} + \dfrac{\lambda}{\sqrt{2\pi\,\omega x}} \cos(\omega x - \tfrac{1}{4}\pi)\cos\omega t$

$$-\frac{\lambda^2}{8\pi}(\omega x + \tfrac{1}{2}\sin(2\omega x - \tfrac{1}{4}\pi)\cos 2\omega t).$$

In the neighborhood of any point $\omega x_0 \gg 1$ we therefore have, to terms of first order in λ, the standing wave

(viii) $(-\gamma_{11} = \gamma_{44}) = -\gamma_{22} = \gamma_{33}$

$$= \frac{\lambda}{\sqrt{2\pi\,\omega x_0}}\cos(x - \tfrac{1}{4})\cos t,$$

where we have only retained terms whose derivatives are of the order $1/\sqrt{\omega x_0}$. The wave $\gamma_{11} = -\gamma_{44}$ (coordinates real!) is spurious, but the wave $\gamma_{22} = -\gamma_{33}$ is a true gravitational wave of the transverse type discussed in Part I.

I believe we are here actually dealing with cylindrical waves, which would account for the presence of the terms $\log x$ in β and γ, and that an examination of terms of order λ, in the rigorous form, in the neighborhood of the axis $x = 0$ will show them to be interpretable as weak cylindrical gravitational waves. On the other hand, the term $\lambda^2 \omega x$ in δ points to trouble at ∞. I have not attempted to determine whether this difficulty is intrinsic or accidental. But this is all beside the present point; we have exhibited here a wave field which is approximately represented at mean distances by plane gravitational waves, and which does not behave in the paradoxical manner predicted by the authors.

(f) Pp. 14, 15. The "incomplete wave." In view of the remarks in (d) above concerning the authors rather awkward normalization of γ, we here seek a more manageable one. As the authors point out, it follows from (39a) that

$$\delta = \phi(x + t) + \psi(x - t),$$

and hence the terms of ds^2 involving the differentials of $\bar{x}_1 = x, \bar{x}_4 = t$ may be written

$$-e^{\phi(x+t)}(dx + dt) \cdot e^{\psi(x-t)}(dx - dt).$$

This suggests the more complete normalization

$$\bar{x} + \bar{t} = \int^{x+t} e^{\phi(z)} dz, \bar{x} - \bar{t} = \int^{x-t} e^{\psi(z)} dz$$

than that adopted by the authors under eq. (30b), involving as it does the introduction of an extraneous function g;
    in these new variables

$$\bar{\delta} = 0, \bar{\beta} = \bar{\beta}(\bar{x} + \bar{t}), \bar{\gamma} = \bar{\gamma}(\bar{x} + \bar{t}).$$

Dropping the bars, the field equations (17)–(21) now reduce to the single equation

$$(ix) \qquad \gamma'' + \frac{1}{4}(\gamma'^2 + \beta'^2) = 0$$

where the prime indicates differentiation with respect to the argument $\xi = x + t$. On writing

$$b = \sqrt{B} = e^{\frac{1}{4}(\gamma+\beta)}, c = \sqrt{C} = e^{\frac{1}{4}(\gamma-\beta)}$$

eq. (ix) becomes

$$(x) \qquad b''/b + c''/c = 0;$$

an examination of the Riemann-Christoffel tensor shows that the space is empty only if b'', c'' both vanish.

    First consider the empty-space case; we may then take as the most general form

$$(xi) \qquad b_0 = 1 + \bar{b}(x + t), c_0 = 1 + \bar{c}(x + t),$$

where $\bar{b}, \bar{c}$ are constants. The pure transverse "weak" field

$$(xii) \qquad \gamma_{22} = -2\bar{b}(x + t), \gamma_{33} = -2\bar{c}(x + t)$$

arising therefrom must of course be spurious - although it is not of the type considered by the authors in Part I. It is, however, transformed away with the aid of the vector

$$(xiii) \qquad \xi_1 = \xi_4 = 0, \qquad \xi_2 = -\bar{b}(x + t)y, \xi = -\bar{c}(x + t)z$$

satisfying $\xi^\mu_{,\alpha\alpha} = 0$.

Since we are interested in plane waves, we attempt to find solutions $b(\xi)$, $c(\xi)$ of eq. (x) involving a parameter λ which are such that on allowing $\lambda \to 0$ both b and $c \to 1$. A decent-behaving class of solutions may be obtained on setting

$$(xiv) \qquad b = 1 + \lambda f(\xi),$$

where $f(\xi)$ is a bounded function of ξ which approaches 0 sufficiently fast as $|\xi| \to \infty$, but is otherwise arbitrary, the remaining function $c(\lambda, \xi)$ is then taken as that particular solution of the linear differential equation

$$(xv) \qquad c'' + \lambda F(\lambda, \xi)c = 0, \quad \text{where} \quad F = f''/(I + \lambda f),$$

which $\to 1$ as $\xi \to +\infty$. Then as $\lambda \to 0$, $c(\lambda, \xi) \to 1$, as required. Admittedly, as $\xi \to -\infty$

$$(xvi) \qquad c(\lambda, \xi) \to \bar{c}_0 + \bar{c}(x + t),$$

where $\bar{c}_0$, $\bar{c}$ are constants depending on the parameter λ in such a way that they approach 1, 0, respectively, as $\lambda \to 0$; I take this to be but a form of the old difficulty with plane gravitational waves - here they would seem to go in plane at $x = +\infty$ and come out distorted at $x = -\infty$; but at any rate the field itself is flat as $|x| \to \infty$ either way.

Now, for sufficiently small λ, the above field is represented by a weak field

$$(xvii) \qquad b = 1 + \lambda f(x + t),$$
$$c = 1 - \lambda[f(x + t) + \text{const.} \cdot (x + t)]$$

to terms of first order in λ; hence

$$(xviii) \qquad -\gamma_{22} = +\gamma_{33} = 2\lambda f(x + t)$$

except for the possible spurious wave arising from the term const $\cdot (x + t)$ in γ_{33}.

Finally, one might ask whether the obvious boundary value problem associated with (xv)—i.e. to find eigenvalues λ such that $c \rightarrow$ const. both as $\xi \rightarrow +\infty$ and $-\infty$ — could be interpreted in the light of the authors' last paragraph on p. 17. I would be inclined to doubt that this is the case, although I have not investigated the matter more thoroughly.

(g) Pp. 16–17. Although I would certainly admit that there may be difficulties connected with the problem of plane gravitational waves, it seems to me that these difficulties are connected with boundary value problems involving space-time as a whole, rather than with anything that may crop up in the analysis of a limited region of space-time. But the results presented by the authors deal with this latter type of problem, most especially in the crucial analysis on pp. 13–15. In view of the comments (e), (f) above, I would maintain that this particular kind of difficulty does not actually arise—although there may be other difficulties in treating infinite plane gravitational waves rigorously, say of the kind stressed by Eddington, Baldwin and Jeffery (P. R. S. 111A, p. 95, 1926), or arising from the behavior of the potentials (even in the classical theory) associated with infinite distributions.

Interviews
and Other New Sources

James Anderson	Hoboken, New Jersey	03/04/95	tape
Peter Bergmann	New York City	03/04/95	notes
Luc Blanchet	Meudon, France	12/10/94	tape
Hermann Bondi	Cambridge, England	7/11/94	tape
Dieter Brill	College Park, Maryland	06/04/95	tape
S. Chandrasekhar	Chicago, Illinois	12/06/95	notes
Fred Cooperstock	Victoria, BC, Canada	26/06/95	tape, correspondence
Thibault Damour	Bures-sur-Yvette, France	11/10/94	tape, correspondence
Nathalie Deruelle	Paris, France	19/10/94	notes
Jürgen Ehlers	Munich, Germany	14/10/94	tape
Joshua Goldberg	Syracuse, New York	10/04/95	notes
Peter Havas	Philadelphia, Pennsylvania	05/04/95	notes, correspondence
Richard Isaacson	Arlington, Virginia	08/04/95	notes
Charles Misner	College Park, Maryland	07/04/95	notes
Ezra T Newman	Pittsburgh, Pennsylvania	11/04/95	notes
Asher Peres			correspondence

Felix Pirani	London, England	25/10/94	tape
Jerzy Plebanski	Mexico City	30/06/95	notes
Adrian Scheidegger			correspondence
Dennis Sciama	Venice, Italy	16/10/94	tape
John Stachel			correspondence
Kip Thorne	Pasadena, California	14/06/95	tape
	Pasadena, California	17/07/95	tape
Andrzej Trautman	Trieste, Italy	17/10/94	tape
Phillip Wallace	Victoria, BC, Canada	26/06/95	notes
Joseph Weber	Irvine, California	20/06/95	tape
John Wheeler	Princeton, New Jersey	04/04/95	notes
Jeffrey Winicour	Pittsburgh, Pennsylvania	11/04/95	notes

Dates are given as day/month/year. Interviews which were recorded and for which a tape is available are indicated. Provided the interviewee is willing, access to the tapes or notes arising from interviews and discussions will be permitted to interested scholars.

NOTES

CHAPTER 1: THE GRAVITATIONAL WAVE ANALOGY

1. For papers that discuss the role which the analogy with electromagnetism played in his thinking, see the papers by Renn and Sauer (2006) and Janssen and Renn (2006). There are many other references, discussing every aspect of the story, that can be followed from those papers.

2. Kuhn calls these problems and their solutions exemplars, and they take over the some of the meanings incorporated in the term *paradigm* as used in his most famous work, *The Structure of Scientific Revolutions.*

3. The ancestor of the General Relativity and Gravitation (GR) series of conferences, and sometimes referred to as GR0. In the transcript of this conference, which was made from a tape recording, certain exchanges by the participants are enclosed in quotation marks, which may indicate direct quotation, whereas other passages are not, which may indicate that the responses were paraphrased. Hence, I have retained the quotation marks when present in the transcript.

CHAPTER 2: THE PREHISTORY OF GRAVITATIONAL WAVES

1. No doubt his acrimonious relationship with John Flamsteed, the Astronomer Royal, over the ownership of the data he chiefly relied upon contributed to his headache.

2. The center of gravity of the Earth-Moon system lies some thousand miles beneath the Earth's surface, still a considerable distance from the center.

3. Although perturbations and increased mass would not now be thought of as, in themselves, dissipative effects, leading to dynamical decay in the sense of loss

of momentum from a system, to Newton and Halley they clearly did. Both seem
to have associated such effects with "loss of motion" and dynamical decay (Kubrin
1995).

4. In general relativity the first relativistic correction to orbital motion occurs
only at the order $(v/c)^2$, and the first nonconservative correction occurs only at
order $(v/c)^5$. Therefore there is no reason to expect this effect to make itself felt
in the Earth-Moon system.

5. That there was some competition to the tidal friction explanation of the
secular acceleration even in the 1950s is shown by the efforts to find part of its
cause instead in a long-term change in the gravitational constant, G (Dicke 1966).

6. This calculation was also made by Lorentz, who was the first to calculate
radiation reaction for a moving charge in the electromagnetic field (Van Lunteren
1991; Damour 1982).

7. Abraham died tragically young, just after he finally secured a professorship
in Germany.

8. The lowest order of radiation present in a field is directly related, in quantum
field theory, to the spin of the mediating particle. A scalar field, which would permit
monopole radiation, would be mediated by particles with no spin. Vector fields,
like electromagnetism, are dominated by dipole radiation and have mediating
particles with spin 1, as photons do. Tensor fields, like general relativity, have only
quadrupole radiation at the lowest order and have mediating particles with spin 2,
which is why gravitons are thought to have this spin.

9. The trace of a tensorial quantity is an ordinary, or scalar, number, obtained
by adding together the diagonal components of the tensor.

CHAPTER 3: THE ORIGINS OF GRAVITATIONAL WAVES

1. These field equations had the Ricci tensor in place of the Einstein tensor; in
other words, the term involving the trace of the Ricci tensor was missing.

2. Electromagnetic theory, on the other hand, is a vector theory; the potential
of the field is described by a four-dimensional vector.

3. As mentioned earlier, the post-Newtonian approximation involves an expansion in two small quantities, one depending on v/c and the other on the
gravitational potential. But when one deals with systems whose motion is con-
trolled by their own gravitational interaction, then the gravitational potential is
directly related to their orbital speed v and so the whole expansion can be written
in terms of v/c if desired.

4. In addition, as we shall see in the next chapter, he had one remaining error
that he did not discover at this time, which meant that the quadrupole formula he
presented in 1918 differs from the real one by a factor of two.

Chapter 4: The Speed of Thought

1. This argument encapsulates the compelling intuitive argument that helped make gravitational waves an accepted phenomenon of modern physics for decades before any experimental evidence in their favor emerged.

2. By describing the energy in the waves as an "analytical fiction," Eddington means that the gravitational wave energy can only be observed when it is absorbed by a physical system, so we should really not speak of energy in the waves, but only of energy exchanged between two systems as a result of the radiation. Later on we shall see how some physicists, such as John Wheeler and Richard Feynman, suggested that the emission of radiation demands that there be an absorber to receive the energy. Also note that Eddington employs a somewhat different accounting system from the one commonly used in describing the equations of motion of binary star systems. In addition he adopts the common practice of counting relativistic order in multiples of v^2/c^2. In the case of a binary star system, and in the terms used in this book, the order of the Laplace term is v/c and the modern term is at v^5/c^5. The reason that Eddington's calculation applies only to "cohesive systems" is that the linearized approximation assumes that the gravitational force is very weak, but a binary star system's motion is completely determined by its own gravitational interaction, so the size of its gravitational force must be of the same order as its velocity, in relativistic units. A spinning rod, on the other hand, may be made to spin as fast as you like, while its gravitational field is vanishingly small.

Chapter 5: Do Gravitational Waves Exist?

1. Although the original version of Einstein and Rosen's paper no longer exists, its original title is referred to in the report by the *Review's* referee (Robertson Papers, Caltech, box 7.12; and EA 19-090), for more of which, see below.

2. The translation from the German is by Diana Buchwald. The emphasis in the letter is Einstein's.

3. Einstein's bibliography to 1949, given by Schilpp (1949), lists no papers by him appearing in the *Physical Review* after 1936, and the index of the *Review* from then until his death refers only to one short note of rebuttal, which was mentioned by Pais (1982, pp. 494–495) in his brief account of the rejection of the Einstein-Rosen paper.

4. Since general relativity is a nonlinear theory, the fact that two potentials (the advanced and retarded) satisfy the field equations does not imply that their linear combination (e.g., half-advanced-plus-half-retarded) would, as it does in electromagnetism. In linearized gravity, however, this obviously does follow.

CHAPTER 6: GRAVITATIONAL WAVES AND THE RENAISSANCE
OF GENERAL RELATIVITY

1. When I speak of Einstein's quadrupole formula I do not, strictly speaking, refer to the equation actually printed in Einstein's paper of 1918 but rather to the corrected version first published by Eddington.

2. See Eisenstaedt, 1993 and references therein for this history.

CHAPTER 7: DEBATING THE ANALOGY

1. Of course, measuring relative displacement of several different pairs of particles is exactly what interferometric gravitational wave detectors such as LIGO do.

2. A 1992 paper by the Canadian physicist Fred Cooperstock contains a counterexample designed to invalidate this thought experiment.

3. It is unclear if "crackpots" is a reference to Weber. It may be that the "crackpots" are those who look for evidence of waves close to the source, where the wave is indistinguishable from the rest of the dynamical field, as Feynman pointed out. Alternatively it may be those who look for distant sources of waves who are crackpots, because of the weakness of the waves.

CHAPTER 8: THE PROBLEM OF MOTION

1. First-order post-Newtonian effects such as the celebrated perihelion shift of Mercury, which appear at order $(v/c)^2$ in the expansions, are not related to gravitational radiation. They are what are known as conservative effects, since they do not change the mechanical energy in the orbiting system. Conservative effects always appear at even order in the post-Newtonian expansions in powers of v/c, while nonconservative effects, including gravitation radiation effects, appear at odd order in the expansions. The lowest order at which nonconservative effects appear in general relativity is $(v/c)^5$, known as post-$2\frac{1}{2}$- Newtonian order.

2. This trick depended on Stoke's theorem, turning volume integrals inside the surface, which did depend on the details of the singularity, into surface integrals on the boundary, which did not.

3. The energy in the wave, as you might expect, falls of as the inverse square of the distance from the source, but it is really the size or amplitude of the wave that determines its detectability.

4. Fock's analogy for this was with "Copernican" versus "Ptolemaic" coordinates. Both could be employed for calculation purposes, but he insisted that the first

must be given a priori status as the correct physical description of the solar system. According to Gorelik (1993, p. 316), the issue of the correctness of Copernicus's choice of coordinates over those of Ptolemy was often cited by those who opposed relativity theory.

Chapter 9: Portrait of the Skeptics

1. Again, while Goldberg clearly recalled some debate on this topic at Chapel Hill, Bryce DeWitt, who organized the meeting, was quite certain, when replying to a presentation by the present author at the meeting in Moscow in 1996, that there was no significant debate on the existence of gravitational waves in the late '50s. On the whole, the conference proceedings, transcribed from a tape recording of the main sessions, tend to bear out Goldberg's recollection. The varying recollections are, however, a signal warning to anyone interested in reconstructing this type of recent history of science.

Chapter 10: On the Verge of Detection

1. Ironically, ocean waves are typically "gravity waves," that is, water waves whose restoring force is supplied by the water's own weight and not by surface tension, as with short-wavelength ripples. It is because of these gravity waves that gravitational waves acquired the slightly unwieldy name they bear. It is a sign of their relatively recent emergence as an important physical phenomenon that the term gravity waves is now increasingly applied to gravitational waves even by physicists.

2. The geon narrowly escaped being called a *kugelblitz*.

3. I'm thinking here of a remark I heard Weber make during a talk at a Pacific Coast gravity meeting in the early to midnineties while presenting his new cross section for bar detectors. He said that he had himself developed the original cross-section model for bar detectors and had been "laughed at then too."

4. I myself have made calculations looking for floating orbits, having realized, while a graduate student, that such things could exist. In my case, the possibility occurred to me only because of a bug in my code which suggested that this had happened. When a fellow student had the same idea, we got together and quickly discovered that the idea was an old one, but no one had ever written a paper announcing that they had NOT found a floating orbit, since scientists often view negative results as unworthy of publication. Since many before us had looked in vain we put into our separate papers a mention of our failed efforts to find one.

CHAPTER 11: THE QUADRUPOLE FORMULA CONTROVERSY

1. A similar viewpoint regarding the absolute insensitivity of bar detectors has been put forward by Lluis Bel (1996).

CHAPTER 12: KEEPING UP WITH THE SPEED OF THOUGHT

1. The super star model was an early model intended to explain the enormous luminosity of quasars.

2. Kepler is quoted in Chandrasekhar's *Truth and Beauty* (1987, p. 66): "Now, it might be asked if this quality of the soul, which does not engage in conceptual thinking and can therefore have no prior knowledge of harmonic relations, should be capable of recognizing what is given in the outward world. . . . To this, I answer that all pure Ideas, or archetypal patterns of harmony, such as we are speaking of, are inherently present in those who are capable of apprehending them. But they are not first received into the mind by a conceptual process, being the product, rather, of a sort of instinctive intuition and innate to those individuals."

3. It's true, of course, that one aspect of the Newton's apple story is that of *observation*. But what is being observed is very commonplace. The importance of the story lies in the intuitive leap by which Newton relates the falling of the apple to the motion of the Moon and asks whether the same agency can be responsible for both.

4. The quoted article continues, "In the study of gravitational waves the chase seems to have become a rout, since no positive experimental results at all are available to justify the copious writings on the subject."

BIBLIOGRAPHY

References such as EA 19-090 refer to documents in the Einstein Archives, Hebrew University of Jerusalem folder 19, document 90.

Anderson, James L. (1980). "New Derivations of the Quadrupole Formula and Balance Equations for Gravitationally Bound Systems." *Physical Review Letters* **45**, 1745–1748.

——— (1987). "Gravitational Radiation Damping in Systems with Compact Components." *Physical Review D* **36**, 2301–2313.

——— (1992). "Why We Use Retarded Potential". *American Journal of Physics* **60**, 465–467.

——— (1995). "Conditions of Motion for Radiating Charged Particles." Preprint, Stevens Institute of Technology, Hoboken, New Jersey.

Armitage, Angus (1966). *Edmond Halley* (Nelson; London).

Baldwin, O. R., and G. B. Jeffery (1926). "The Relativity Theory of Plane Waves." *Proceedings of the Royal Society of London, series A* **111**, 95–104.

Beck, Guido (1925). "Zur Theorie binärer Gravitationsfelder." *Zeitschrift für Physik* **33**, 713–728.

Bel, Lluis (1996). "Static Elastic Deformations in General Relativity." Electronic preprint (gr-qc/9609045), http://xxx.lanl.gov (accessed June 19, 2006).

Bel, Lluis, Thibault Damour, Nathalic Deruelle, Jesus Ibanez, and Jesus Martin (1981). "Poincaré-Invariant Gravitational Field and Equations of Motion of Two Pointlike Objects: The Postlinear Approximation of General Relativity." *General Relativity and Gravitation* **13**, 963–1004.

Bergmann, Peter (1942). *Introduction to the Theory of Relativity* (Prentice-Hall, New Jersey).

Bergmann, Peter (1968). *The Riddle of Gravitation* (Scribner, New York).

Bertotti, Bruno, and Jerzy Plebanski (1960). "Theory of Gravitational Perturbations in the Fast Motion Approximation." *Annals of Physics* **11**, 169–200.

Bondi, Hermann (1957). "Plane Gravitational Waves in General Relativity." *Nature* **179**, 1072–1073.

——— (1964). "Radiation from an Isolated System." In *Relativistic Theories of Gravity*, Proceedings of the Warsaw Conference, July 25–31, 1962, ed. Leopold Infeld, pp. 120–121 (Gauthier-Villiers, Paris).

——— (1970). "General Relativity as an Open Theory." In *Physics, Logic and History*, ed. W. Yourgrau and A. D. Breck (Plenum, New York).

——— (1987). "Gravitating toward Wave Theory." *The Scientist* **1**, 17.

——— (1990). *Science, Churchill, and Me* (Pergamon, New York).

Bondi, Hermann, Felix A. E. Pirani, and Ivor Robinson (1959). "Gravitational Waves in General Relativity III. Exact Plane Waves." *Proceedings of the Royal Society of London, Series A* **251**, 519–533.

Bondi, Hermann, M. G. J. van der Burg, and A. W. K. Metzner (1962). "Gravitational Waves in General Relativity VII: Waves from Axi-Symmetric Isolated Systems." *Proceedings of the Royal Society of London, Series A* **269**, 21–52.

Bonnor, W. B. (1963). "Gravitational Waves." *British Journal of Applied Physics* **14**, 555–562.

Born, Max (1971). Letter no. 71. *The Born-Einstein Letters* (Walker, New York).

Bourdieu, Pierre (1975). "The Specificity of the Scientific Field and the Social Conditions of the Progress of Reason." *Social Science Information* **14**, 19–47.

Brans, C., and R. H. Dicke (1961). "Mach's Principle and a Relativistic Theory of Gravitation." *Physical Review* **124**, 925–935.

Brill, Dieter R. (1959). "On the Positive Definite Mass of the Bondi-Weber-Wheeler Time-Symmetric Gravitational Waves." *Annals of Physics* **7**, 466–483.

Brill, Dieter R., and James B. Hartle (1964). "Method of the Self-Consistent Field in General Relativity and Its Application to the Gravitational Geon." *Physical Review* **135**, B271–B278.

Brosche, P. (1977). "Kant und die Gezeitenreibung." *Die Sterne* **53**, 114–117.

Burke, William (1969). "The Coupling of Gravitational Radiation to Nonrelativistic Sources." Ph.D. diss., California Institute of Technology.

——— (1971). "Gravitational Radiation Damping of Slowly Moving Systems Calculated Using Matched Asymptotic Expansions." *Journal of Math. Physics* **12**, 402–418.

Burke, William, and Kip S. Thorne (1970). "Gravitational Radiation Damping." In *Relativity*, ed. Moshe Carmeli, Stuart I. Fickler, and Louis Witten, pp. 209–228 (Plenum Press, New York).

Cameron, A.G.W. (1963). *Interstellar Communications* (W. A. Benjamin, New York).

Cattani, Carlo, and Michelangelo De Maria (1993). "Conservation Laws and Gravitational Waves in General Relativity (1915–1918)." In Earman, Janssen, and Norton (1993), pp. 63–87.

Chandrasekhar, Subrahmanyan (1965). "The post-Newtonian Equations of Hydrodynamics in General Relativity." *Astrophysical Journal* **142**, 1488–1512.

———— (1987). *Truth and Beauty: Aesthetics and Motivations in Science* (Chicago, University Press).

Chandrasekhar, Subrahmanyan, and F. P. Esposito (1970). "The $2\frac{1}{2}$-post-Newtonian Equations of Hydrodynamics and Radiation Reaction in General Relativity." *Astrophysical Journal* **160**, 153–179.

Clifford, William Kingdom (1876). "On the Space-Theory of Matter." *Proceedings of the Cambridge Philosophical Society* **2**, 157–158.

Cohen, I. Bernard (1975). *Isaac Newton's Theory of the Moon* (Dawson, Folkestone, England).

Collins, Harry M. (1994). "A Strong Confirmation of the Experimenters' Regress." *Studies in History and Philosophy of Science Part A* **25**, 493–503.

———— (2001). "Tacit Knowledge, Trust and the Q of Sapphire." *Social Studies of Science* **31**, 71–85.

———— (2004). *Gravity's Shadow* (University of Chicago Press, Chicago).

Cooperstock, Fred I. (1967). "Energy Transfer via Gravitational Radiation in the Quasistellar Sources." *Physical Review* **163**, 1368–1373.

———— (1974). "Axially Symmetric Two-Body Problem in General Relativity." *Physical Review D* **10**, 3171–3180.

———— (1992). "Energy Localization in General Relativity: A New Hypothesis." *Foundations of Physics* **22**, 1011–1024.

Cooperstock, Fred I., V. Faraoni, and G.P. Perry (1995). "Can a Gravitational Geon Exist in General Relativity?" *Modern Physics Letters A* **10**, 359–365.

Cooperstock, Fred I., and D. W. Hobill (1982). "Gravitational Radiation and the Motion of Bodies in General Relativity." *General Relativity and Gravitation* **14**, 361–378.

Damour, Thibault (1982). "Gravitational Radiation and the Motion of Compact Bodies." In *Rayonnement Gravitationelle*, ed. N. Deruelle and T. Piran, p. 59–144 (North Holland, Amsterdam).

———— (1983). "Gravitational Radiation Reaction in the Binary Pulsar and the Quadrupole-Formula Controversy." *Physical Review Letters* **51**, 1019–1021.

Damour, Thibault (1987). "An Introduction to the Theory of Gravitational Radiation." *Gravitational in Astrophysics*, ed. Brandon Carter and James B. Hartle, pp. 3–62 (Plenum Press, New York).

Damour, Thibault, and R. Ruffini (1974). "Sur certaines vérifications nouvelles de la Relativité générale rendues possibles par la découverte d'un pulsar membre d'un système binaire." *Comptes Rendu de l'Academie des Sciences de Paris, Series A* **279**, 971–973.

Davies, P.C.W. (1980). *The Search for Gravity Waves* (Cambridge University Press, Cambridge).

De Sitter, Willem (1916). "On Einstein's Theory of Gravitation and its Astronomical Consequences." *Monthly Notices of the Royal Astronomical Society* **77**, 155–184.

DeWitt, Cécile M., ed. (1957). *Conference on the Role of Gravitation in Physics*. Proceedings of the conference at Chapel Hill, North Carolina, January 18–23, 1957. WADC (Wright Air Development Center) technical report 57-216, (United States Air Force, Wright-Patterson Air Force Base, Ohio). A supplement with an expanded synopsis of Feynman's remarks was also distributed to participants, a copy of which can be found in the Feynman Papers at the California Institute of Technology, Box 91, Folder 2.

Dicke, R. H. (1966). "The Secular Acceleration of the Earth's Rotation and Cosmology." In *The Earth-Moon System*, ed. B. G. Marsden and A.G.W. Cameron (Plenum, New York).

Dixon, W. G. (1979). "Extended Bodies in General Relativity: Their Description and Motion." In Ehlers (1979).

Droste, J. (1917). "The Field of a Single Centre in Einstein's Theory of Gravitation and the Motion of a Particle in That Field." *Proc. Acad. Sci. Amsterdam* **19**, 197–215.

Dyson, Freeman (1963). "Gravitational Machines." In *Interstellar Communications*, ed. A.G.W. Cameron, pp. 115–120 (W.A. Benjamin, New York).

Earman, John (1989). *World Enough and Space-Time. Absolute versus Relational Theories of Space and Time* (MIT Press, Cambridge, Mass).

Earman, John, and Jean Eisenstaedt (1999). "Einstein and Singularities." *Studies in History and Philosophy of Modern Physics* **30B**, 185–235.

Earman, John, and Michel Janssen (1993). "Einstein's Explanation of the Motion of Mercury's Perihelion." In Earman, Janssen, and Norton (1993), pp. 129–172.

Earman, John, Michel Janssen, and John D. Norton, eds. (1993). *The Attraction of Gravitation: New Studies in the History of General Relativity* (Birkhäuser, Boston).

Eddington, Arthur Stanley (1922). "The Propagation of Gravitational Waves." *Proceedings of the Royal Society of London, series A* **102**, 268–282.

—— (1923a). *The Mathematical Theory of Relativity* (Cambridge University Press, Cambridge).

—— (1923b). "The Spontaneous Loss of Energy of a Spinning Rod according to the Relativity Theory." *Philosophical Magazine* **46**, 1112–1117.

Edge, David, and Michael J. Mulkay (1976). *Astronomy Transformed: The Emergence of Radio Astronomy in Britain* (New York, Wiley).

Ehlers, Jürgen, ed. (1979). *Isolated Gravitating Systems in General Relativity*. Proceedings of the International School of Physics "Enrico Fermi," Course 67, Varenna, Italy, June 28-July 10, 1976 (North Holland, Amsterdam).

—— (1980). "Isolated Systems in General Relativity." In *Ninth Texas Symposium on Relativistic Astrophysics* ed J. Ehlers, J. J. Perry, and M. Walker, pp. 279–294 (New York Academy of Science, New York).

—— (1987). "Folklore in Relativity and What Is Really Known." In *General Relativity and Gravitation: Proceedings of the 11th International Conference on General Relativity and Gravitation*, Stockholm, July 6–12, 1986, ed. M.A.H. MacCallum (Cambridge University Press, Cambridge).

Ehlers, Jürgen, Arnold Rosenblum, Joshua N. Goldberg, and Peter Havas (1976). "Comments on Gravitational Radiation Damping and Energy Loss in Binary Systems." *The Astrophysical Journal* **208**, L77–L81.

Einstein, Albert (1915). "Die Feldgleichungen der Gravitation." *Königlich Preussische Akademic der Wissenschaften Zu Berlin, Sitzungsberichte* (1915), pp. 844–847.

—— (1916). "Näherungsweise Integration der Feldgleichungen der Gravitation." *Königlich Preussische Akademie der Wissenschaften Zu Berlin, Sitzungsberichte*, pp. 688–696.

—— (1918). "Über Gravitationswellen." *Königlich Preussische Akademie der Wissenschaften Zu Berlin, Sitzungsberichte*, pp. 154–167.

—— (1998). *The Collected Papers of Albert Einstein*, Vol. 8, ed. Robert Schulmann, A. J. Kox, Michel Janssen, and József Illy (Princeton University Press, Princeton, New Jersey.) Translations by Ann M. Hentschel in accompanying volume.

—— (2002). "Excerpt from Lecture Notes for Course on General Relativity." In *The Collected Papers of Albert Einstein*, Vol. 7, ed. Michel Janssen, Robert Schulmann, József Illy, Christoph Lehner, and Diana Kormos Buchwald, pp. 185–189 (Princeton University Press, Princeton, New Jersey).

Einstein, Albert, and Jakob Grommer (1927). "Allgemeine Relativitätstheorie und Bewegungsgesetz." *Königlich Preussische Akademie der Wissenschaften Zu Berlin, Sitzungsberichte*, pp. 2–13.

Einstein, Albert, Leopold Infeld, and Banesh Hoffmann (1938). "The Gravitational Equations and the Problem of Motion." *Annals of Mathematics* **39**, 65–100.

Einstein, Albert, and Walter Ritz (1909). "Zum gegenwärtigen Stand des Strahlungsproblems." *Physikalische Zeitschrift* **10**, 323–324. Translated by Anna Beck in *Collected Papers of Albert Einstein,* Vol. 2, *The Swiss Years: Writings 1900–1909,* p. 376 (Princeton University Press, Princeton, New Jersey).

Einstein, Albert, and Nathan Rosen (1937). "On Gravitational Waves." *Journal of the Franklin Institute* **223**, 43–54.

Eisenstaedt, Jean (1986a). "La relativité générale a l'étiage: 1925–1955." *Archive for the History of Exact Sciences* **35**, 115–185.

——— (1986b). "The Low Water Mark of General Relativity, 1925–1955." In *Einstein and the History of General Relativity,* ed. D. Howard and J. Stachel, pp. 277–292 (Birkhäuser, Boston).

——— (1993). "Lemaître and the Schwarzschild Solution." In Earman, Janssen, and Norton (1993).

——— (2006). *The Curious History of Relativity: How Einstein's Theory of Gravity was Lost and Found Again* (Princeton University Press, Princeton, New Jersey).

Felber, H.-J. (1974). "Kant Beitrag zur Frage der Verzögerung der Erdrotation." *Die Sterne* **50**, 82–90.

Feyerabend, Paul K. (1988). *Against Method* (Verso, London).

Feynman, Richard P. (1984). *Surely You're Joking Mr. Feynman: Adventures of a Curious Character* (Norton, New York).

——— (1995). *Feynman Lectures on Gravitation* (Addison-Wesley, Reading, Mass).

——— (2005). *Perfectly Reasonable Deviations from the Beaten Path: The Collected Letters of Richard P. Feynman.* Ed. Michelle Feynman (Basic Books, New York).

Feynman, Richard P., and Ralph Leighton (1989). *What Do You Care What Other People Think? Further Adventures of a Curious Character* (Norton, New York). (Bantam, New York).

Fock, Vladimir A. (1959). *The Theory of Space, Time and Gravitation.* Trans. N. Kemmer (Pergamon, New York). First English edition.

Fowler, William A. (1964). "Massive Stars, Relativistic Polytropes and Gravitational Radiation: Gravitational Waves as Trigger for Radio Galaxy Emissions." *Reviews of Modern Physics* **36**, 545–555.

Franklin, Allan (1994). "How to Avoid the Experimenters' Regress." *Studies in History and Philosophy of Science Part A* **25**, 463–491.

Galison, Peter (1987). *How Experiments End* (University of Chicago Press, Chicago).

——— (1995). "Theory Bound and Unbound: Superstrings and Experiment." In *Laws of Nature: Essays on the Philosophical, Scientific and Historical Dimensions,* ed. Friedel Weinert, pp. 369–408 (Walter de Gruyter, Berlin).

Gold, Thomas (1967). *The Nature of Time* (Cornell Universty Press, Ithaca, New York).

Goldberg, Joshua N. (1955). "Gravitational Radiation." *Physical Review* **99**, 1873–1883.

———— (1966). "Gravitation." In *The Encyclopedia of Physics*, ed. Robert M. Bescanson, p. 300 (Reinhold, New York).

———— (1974). "Gravitation." In *The Encyclopedia of Physics*, ed. Robert M. Bescanson, p. 396 (Reinhold, New York).

———— (1993). "US Air Force Support of General Relativity: 1956–1972." In Kox and Eisenstaedt (1992), pp. 89–102.

Gorelik, Gennady (1993). "Fock: Philosophy of Gravity and Gravity of Philosophy." In Earman, Janssen, and Norton (1993), pp. 308–331.

Graham, Loren R. (1987). *Science, Philosophy and Human Behavior in the Soviet Union* (Columbia University Press, New York).

Graves, Robert (1966). *The White Goddess* (Farrar, Straus and Giroux, New York).

Hall, Karl (2005). " 'Think Less about Foundations': A Short Course on Landau and Lifshitz's Course of Theoretical Physics." In *Pedagogy and the Practice of Science: Historical and Contemporary Perspectives*, ed. David Kaiser, pp. 253–256 (MIT Press, Cambridge).

Havas, Peter (1957). "Radiation Damping in General Relativity." *Physical Review* **108**, 1351–1352.

———— (1974). "Equations of Motion, Radiation Reaction, and Gravitational Radiation." In *Ondes et Radiation Gravitationelles*. Proceedings of the meeting at Paris, June 18–22, 1973 pp. 383–392. Colloques internationaux du centre national de la recherche scientifique, No. 220, (Editions du Centre national de la recherche scientifique, Paris).

———— (1979). "Equations of Motion and Radiation Reaction in the Special and General Theory of Relativity." In *Isolated Gravitating Systems in General Relativity*. ed. Jürgen Ehlers, pp. 74–155 (North Holland, Amsterdam).

———— (1989). "The Early History of the Problem of Motion in General Relativity." In *Einstein and the History of General Relativity*, ed. D. Howard and J. Stachel, pp. 234–276 (Boston, Birkhäuser).

———— (1993). "The Two-Body Problem and the Einstein-Silberstein Controversy." In Earman, Janssen, and Norton (1993), pp. 88–125.

———— (1995). "The Life and Work of Guido Beck: The European Years: 1903–1943." *An. Acad. bras. Ci.*, **67** 11–36.

Havas, Peter, and Joshua N. Goldberg (1962). "Lorentz-Invariant Equations of Motion of Point Masses in the General Theory of Relativity." *Physical Review* **128**, 398–414.

Hesse, Mary B. (1966). *Models and Analogies in Science* (University of Notre Dame Press, South Bend, Indiana).

Hoffmann, Banesh (1972). *Albert Einstein, Creator and Rebel* (Viking, New York).

Hu, Ning (1947). "Radiation Damping in the General Theory of Relativity." *Proceedings of the Royal Irish Academy* **51A**, 87–111.

———— (1982). "Radiation Damping Forces of Binary Stars Due to Emission of Gravitational Waves." In *Proceedings of the Second Marcel Grossmana Meeting on General Relativity*. Organized and held at the International Center for Theoretical Physics, Trieste, July 5–11, 1979, ed. Remo Ruffini, pp. 717–726, (North Holland, Amsterdam).

Hulse, R. A., and J. H. Taylor (1975). "Discovery of a Pulsar in a Binary System." *Astrophysical Journal* **195**, L51–L53.

Infeld, Leopold (1941). *Quest—The Evolution of a Physicist* (Gollancz, London).

————, ed. (1964). *Conférence internationale sur les théories relativistes de la gravitation*. Proceedings of a conference in Warsaw and Jablonna July 25–31, 1962 (Gauthier-Villars, Paris).

———— (1978). *Why I Left Canada*. With an introduction by Lewis Pyenson (McGill-Queen's University Press, Montreal).

Infeld, Leopold, and Róza Michalska-Trautman (1969). "The Two-Body Problem and Gravitational Radiation." *Annals of Physics* **55**, 561–575.

Infeld, Leopold, and Jerzy Plebanski (1960). *Motion and Relativity* (Pergamon, New York).

Infeld, Leopold, and Adrian E. Scheidegger (1951). "Radiation and Gravitational Equations of Motion." *Canadian Journal of Mathematics* **3**, 195–207.

Infeld, Leopold, and Alfred Schild (1949). "On the Motion of Test Particles in General Relativity." *Reviews of Modern Physics* **21**, 408–413.

Infeld, Leopold, and Philip Wallace (1940). "The Equations of Motion in Electrodynamics." *Physical Review* **57**, 797–806.

Isaacson, Richard (1968). "Gravitational Radiation in the Limit of High Frequency II: Nonlinear Terms and the Effective Stress Tensor." *Physical Review* **166**, 1272–1280.

Jaffe, Arthur, and Frank Quinn (1993). " 'Theoretical Mathematics': Toward a Cultural Synthesis of Mathematics and Theoretical Physics." *Bulletin of the American Mathematical Society* **29**, 1–13.

Janssen, Michel (2006). "What Did Einstein Know and When Did He Know It?: A Besso Memo Dated August 1913." In vol. 1 of *The Genesis of General Relativity: Documents and Interpretation*, ed. Jürgen Renn et al. (Springer, Dordrecht).

Janssen, Michel, and Jürgen Renn (2006). "Untying the Knot: How Einstein Found His Way Back to the Field Equations Discarded in the Zurich Notebook."

In vol. 3 of *The Genesis of General Relativity: Documents and Interpretation*, ed. Jürgen Renn et al. (Springer, Dordrecht).

Jungnickel, Christa, and Russel McCormmach (1986). *Intellectual Mastery of Nature*. Vol. 2, *The Now Mighty Theoretical Physics, 1870–1925* (University of Chicago Press, Chicago).

Kaiser, David (2000). "Roger Babson and the Rediscovery of General Relativity." In *Making Theory: Producing Theory and Theorists in Postwar America*. Ph.D. diss., Harvard University.

——— (2005). "The Atomic Secret in Red Hands? American Suspicions of Theoretical Physicists in the Early Cold War." *Representations* **90**, 28–60.

——— (2006). "Whose Mass Is It Anyway? Particle Cosmology and the Objects of Theory." *Social Studies of Science*, forthcoming.

Katzir, Shaul (2005). "Poincaré's Relativistic Theory of Gravitation." In *The Universe of General Relativity*, ed. A. J. Kox and Jean Eisenstaedt, pp. 15–37 (Birkhauser, Boston).

Kennefick, Daniel (2000). "Star Crushing: Theoretical Practice and the Theoreticians' Regress." *Social Studies of Science* **30**, 5–40.

——— (2005). "Einstein and the Problem of Motion: A Small Clue." In *The Universe of General Relativity*, ed. A. J. Kox and Jean Eisenstaedt, pp. 109–124 (Birkhauser, Boston).

Kerr, Roy P. (1959). "On the Lorentz-Invariant Approximation Method in General Relativity. III. The Einstein-Maxwell Field." *Nuovo Cimento* **13**, 673–89.

Kevles, Daniel (1977). *The Physicists* (Knopf, New York).

Kox, Anne J., and Jean Eisenstaedt (1992). *Studies in the History of General Relativity* (Birkhauser, Boston).

Kragh, Helge (1996). *Cosmology and Controversy: The Historical Development of Two Theories of the Universe* (Princeton University Press, Princeton, New Jersey).

Kubrin, David (1995). "Newton and the Cyclical Cosmos." In *Newton*, ed. I. B. Cohen and R. S. Westfall, pp. 281–296 (Norton, New York).

Kuhn, Thomas (1977). "Second Thoughts on Paradigms." In *The Essential Tension* pp. 291–319 (University of Chicago Press, Chicago).

Landau, Lev D., and Evgenii M. Lifshitz (1951) *Classical Theory of Fields* (Addison-Wesley, Cambridge, Mass).

Laplace, Pierre-Simon de (1776). "Sur le Principe de la Gravitation Universelle." Reprinted in *Oeuvres complètes de Laplace*, vol. 8, pp. 201–275 (Gauthier-Villars, Paris, 1891).

Laplace, Pierre-Simon de (1825). *Traité de Mécanique Céleste*, vol. 4, book 10, chapter 7, section 22. Reprinted in *Oeuvres complètes de Laplace* (Gauthier-Villars, Paris, 1891).

Lichnerowicz, André (1955). *Théories relativistes de la gravitation et de l'electromagnetisme* (Masson, Paris).

———— (1993). "Mathematics and General Relativity: A Recollection." In Kox and Eisenstaedt (1992), pp. 103–108.

Lightman, Alan, and Roberta Brawer (1990). *Origins* (Harvard University Press, Cambridge, Mass).

Lord, Alfred B. (2000). *The Singer of Tales* (Harvard University Press, Cambridge, Mass.)

Martinez, Alberto A. (2001). The Neglected Science of Motion: The Kinematic Origins of Relativity. Ph. D. diss. University of Minnesota.

McVittie, G. C. (1955). "Gravitational Waves and One-Dimensional Einsteinian Gas Dynamics." *Journal of Rational Mechanics and Analysis* 4, 201–220.

Mercier, André (1993). "General Relativity at the Turning Point of Its Renewal." In Kox and Eisenstaedt (1992), pp. 109–121.

Mercier, André, and M. Kervaire, eds. (1956). *Jubilee of Relativity Theory*, Proceedings of the anniversary conference at Bern, July 11–16, 1955 (Birkhäuser-Verlag, Basel)

Miller, Arthur (2005). *Empire of the Stars: Obsession, Friendship and Betrayal in the Quest for Black Holes* (Houghton Mifflin, Boston).

Misner, Charles W. (1972). "Interpretion of Gravitational-Wave Observations." *Physical Review Letters* 28, 994–997.

Misner, Charles W., Kip S. Thorne, and John Archibald Wheeler (1973). *Gravitation* (Freeman, San Francisco).

Narasimha, Roddam (2004). "Divide, Conquer and Unify." *Nature* 432, 807.

Newman, Ezra T., and Roger Penrose (1962). "An Approach to Gravitational Radiation by a Method of Spin Co-efficients." *J. Math. Phys.* 3, 566–578.

Nier, A. O. C., and J. H. Van Vleck (1975). "John Torrance Tate." In *Biographical Memoirs*, Vol. 47, pp. 461–484 (National Academies Press, Washington, DC).

North, John D. (1981). "Science and Analogy." In *On Scientific Discovery*, ed. M. D. Grmek, R. S. Cohen, and G. Cimino (Reidel, Dordrecht).

Page, Leigh (1924). "Advanced Potentials and Their Application to Atomic Models." *Physical Review* 24, 296–305.

Pais, Abraham (1982). *Subtle Is the Lord . . . : The Science and Life of Albert Einstein* (Clarendon, Oxford).

Parker, Sybil P. (1983). *McGraw-Hill Encyclopedia of Physics* (McGraw-Hill, New York).

Penrose, Roger (1964). "Conformal Treatment of Infinity." In *Relativity, Groups and Topology*, ed. C. DeWitt and B. S. DeWitt, pp. 563–584 (Gordon and Breach, New York).

—— (1968). "Structure of Spacetime." In *Battelle Rencontres: 1967 Lectures in Mathematics and Physics*, ed. B. S. DeWitt and J. A. Wheeler, pp. 121–235 (Benjamin, New York).

—— (1969). "Gravitational Collapse: The Role of General Relativity." *Revista del Nuovo Cimento* 1, 252–276.

Penrose, Roger, Ivor Robinson, and Jacek Tafel (1997). "Andrzej Mariusz Trautman." *Classical and Quantum Gravity* 14, A1–A8.

Peres, Asher (1959a). "Gravitational Motion and Radiation I." *Nuovo Cimento* 11, 617–627.

—— (1959b). "Gravitational Motion and Radiation II." *Nuovo Cimento* 11, 644–655.

—— (1959c). "On Gravitational Radiation." *Nuovo Cimento* 13, 670.

—— (1960). "Gravitational Radiation." *Nuovo Cimento* 15, 351–369.

Peters, P. C., and Jon Mathews (1963). "Gravitational Radiation from Point Masses in a Keplerian Orbit." *Physical Review* 131, 435–440.

Peterson, Ivars (1993). *Newton's Clock* (Freeman, New York).

Pinch, Trevor (1977). "What Does a Proof Do if It Does Not Prove?" In *The Social Production of Scientific Knowledge* ed. E. Mendelsohn, P. Weingart, and R. Whitley, pp. 171–215 (Dordrecht-Holland, Boston).

Pirani, Felix A. E. (1955). Review of McVittie (1955). *Mathematical Reviews* 16, 1165.

—— (1962). "Gravitational Radiation." In *Gravitation*, ed. Louis Witten, pp. 199–226 (Wiley, New York).

Poincaré, Henri (1908). "La dynamique de l'électron." *Revue générale des sciences pures et appliqués* 19, 386–402 (1908). Reprinted in *Oeuvres de Henri Poincaré*, vol. 9, pp. 551–586. Translated by Francis Maitland as "The New Mechanics," in *Science and Method*, pp. 199–250 (Dover, New York, 1952).

Preskill, John, and Kip S. Thorne (1996). Forward to Feynman 1995.

Press, William H., and Saul A. Teukolsky (1972). "Floating Orbits, Superradiant Scattering and the Black-hole Bomb." *Nature* 238, 211–212.

Pyenson, Lewis (1978). Introduction to Infeld (1978).

Rees, Martin, Remo, Ruffini, and John Archibald Wheeler (1974). *Black Holes, Gravitational Waves and Cosmology: An Introduction to Current Research* (Gordon and Breach, New York).

Renn, Jürgen (2006). "The Summit Almost Scaled: Max Abraham as a Pioneer of a Relativistic Theory of Gravitation." In *The Genesis of General Relativity: Documents and Interpretation*. Vol. 3, ed. Jürgen Renn et al. (Kluwer, Dordrecht).

Renn, Jürgen, and Tilman Sauer (2006). "Pathway Out of Theoretical Physics." In *The Genesis of General Relativity: Documents and Interpretation*. Vol. 3, ed. Jürgen Renn et al. (Kluwer, Dordrecht).

Reuben, William A. (1955). *The Atom Spy Hoax* (Action Books, New York, 1955).

Ritz, Walter (1908). "Recherches Critiques sur l'Electrodynamique Générale." *Annales de Chemie et de Physique* **13**, 145–275. Trans. by R. S. Fitzius. *Critical Researches on General Electrodynamics,* (self published by the translator, 1980).

Robertson, Howard Percy (1938). *Annals of Mathematics* **39**, 101.

Robinson, Ivor, Alfred Schild, and E. L. Schucking, eds. (1965). *Quasi-stellar Sources and Gravitational Collapse, Including the Proceedings of the First Texas Symposium on Relativistic Astrophysics* (University of Chicago Press, Chicago).

Robinson, Ivor, and Andrzej Trautman (1960). "Spherical Gravitational Waves." *Physical Review Letters* **4**, 431–432.

Rosen, Nathan (1937). "Plane Polarized Waves in the General Theory of Relativity." *Physikalische Zeitschrift der Sowjetunion* **12**, 366–372.

———— (1955). "On Cylindrical Gravitational Waves." In *Jubilee of Relativity Theory*. Proceedings of the anniversary conference at Bern, July 11–16, 1955, pp. 171–175 (Birkhäuser-Verlag, Basel).

———— (1958). "Energy and Momentum of Cylindrical Gravitational Waves." *Physical Review* **110**, 291–292.

———— (1964). "Energy and Gravitational Waves in Bi-Metric Relativity Theory." In *Atti del Convegno Sulla Relatività Generale: Problemi Dell'Energia e Onde Gravitazionale*. Firenze, September 9–12, 1964 (Comitato Nazionale per le Manifestazioni Celebrative, Rome).

———— (1979). "Does Gravitational Radiation Exist?" *General Relativity and Gravitation* **10**, 351–364.

Rosen, Nathan, and K. S. Virbhadra (1993). "Energy and Momentum of Cylindrical Gravitational Waves." *General Relativity and Gravitation* **25**, 429–433.

Rosenblum, Arnold (1981). "Gravitational Radiation Energy Loss in Scattering Problems and the Einstein Quadrupole Formula." *Physics Letters* **81A**, 1–4.

———— (1982). "The Third Order Equations of Motion in the Fast Motion Approach in General Relativity." *Physics Letters* **93A**, 11–14.

Roseveare, N. T. (1982). *Mercury's Perihelion from Le Verrier to Einstein* (Clarendon Press, Oxford).

Sachs, Rainer K. (1962). "Gravitational Waves in General Relativity VIII: Waves in Asymptotically Flat Space-Time." *Proceedings of the Royal Society of London, Series A* **270**, 103–126.

———— (1964). "Gravitational Radiation." In *Relativity, Groups and Topology*, ed. C. DeWitt and B. S. DeWitt, pp. 521–562 (Gordon and Breach, New York).

Scheidegger, Adrian E. (1951). "Gravitational Transverse-Transverse Waves." *Physical Review* **82**, 883–885.

——— (1953). "Gravitational Motion." *Reviews of Modern Physics* **25**, 451–468.

Schilpp, Paul Arthur (1949). *Albert Einstein, Philosopher-Scientist* (Library of Living Philosophers, Evanston, Illinois).

Schwarzschild, Karl (1916). "Über das Gravitationsfeld eines Massenpunktes nach der Einsteinschen Theorie." *Sitzber. Deut. Akad. Wiss. Berlin, Kl. Math.-Phys. Tech.* 424–434.

Schweber, Silvan S. (1994). *QED and the Men Who Made It* (Princeton University Press, Princeton, New Jersey).

Schwimming, Rainer (1980). "On the History of the Theoretical Discovery of the Plane Gravitational Waves." (preprint, Leipzig)

Smith, Stanley F., and Peter Havas (1965). "Effects of Gravitational Radiation Reaction in the General Relativistic Two-Body Problem by a Lorentz-Invariant Method." *Physical Review* **138**, B495–B508.

Sobel, Dava (1995). *Longitude* (Penguin Books, Middlesex, England).

Stachel, John (1959). "Energy Flow in Cylindrical Gravitational Waves." Master's thesis, Stevens Institute of Technology.

——— (2002). "'The Relations Between Things' versus 'The Things Between Relations': The Deeper Meaning of the Hole Argument." In *Reading Natural Philosophy: Essays in the History and Philosophy of Science and Mathematics*, ed. David B. Malament pp. 231–266 (Open Court, Chicago and La Salle).

Stanley, Matthew (2003). "An Expedition to Heal the Wounds of War: The 1919 Eclipse and Eddington as a Quaker Adventurer." *Isis* **94**, 57–89.

Taylor, J. H., and P. M. McCulloch (1980). "Evidence for the Existence of Gravitational Radiation from Measurements of the Binary Pulsar PSR 1913+16." In *Proceedings of the Ninth Texas Symposium on Relativistic Astrophysics*, ed. Jürgen Ehlers, Judith Perry, and Martin Walker, pp. 442–446 (New York Academy of Sciences, New York).

Tetrode, Hugo M. (1922). *Zeitschrift für Physik* **10**, 317–325. The quoted translation is by J. Dorling.

Thorne, Kip S. (1969). "Nonradial Pulsation of General-Relativistic Stellar Models, III: Analytic and Numerical Results for Neutron Stars." *Astrophysical Journal* **158**, 1–16.

——— (1980). "Multipole Expansions of Gravitational Radiation." *Reviews of Modern Physics* **52**, 299–339.

——— (1989). Unpublished manuscript on gravitational radiation, including a historical review of the radiation reaction controversy.

——— (1994). *Black Holes and Time Warps: Einstein's Outrageous Legacy* (Norton, New York).

Thorne, Kip S. and Sándor J. Kovács (1974). "The Generation of Gravitational Waves I: Weak-Field Sources." *Astrophysical Journal* **200**, 245–262.

Trautman, Andrzej (1958a). "Radiation and Boundary Conditions in the Theory of Gravitation." *Bulletin de l'Academie Polonaise des Sciences, Series des Sciences mathématiques* **6**, 407–412.

——— (1958b). "On Gravitational Radiation Damping." *Bulletin de l'Académie Polonaise des Sciences, Series des Sciences mathématiques* **6**, 627–633.

Trautman, Andrzej, Felix Pirani, and Hermann Bondi (1965). *Lectures in General Relativity* (Prentice-Hall, New Jersey).

Van Lunteren, Frans Herbert (1991). *Framing Hypotheses: Conceptions of Gravity in the 18th and 19th Centuries*. Ph.D. diss. Rijksuniversiteit te Utrecht.

Wali, Kameshwar C. (1992). *Chandra: A Biography of S. Chandrasekhar* (University of Chicago Press, Chicago).

Walker, Martin, and Clifford M. Will (1980). "The Approximation of Radiative Effects in Relativistic Gravity: Gravitational Radiation Reaction and Energy Loss in Nearly Newtonian Systems." *Astrophysical Journal* **242**, L129–L133.

Wallace, Phillip (1993). "The Beginnings of Theoretical Physics in Canada." *Physics in Canada* (November), 358–364.

Warwick, Andrew (2005). *Masters of Theory: Cambridge and the Rise of Mathematical Physics* (University of Chicago Press, Chicago).

Weber, Joseph (1960). "Detection and Generation of Gravitational Waves." *Physical Review* **117**, 306–313.

——— (1961). *General Relativity and Gravitational Waves* (Interscience, New York).

——— (1969). "Evidence for Discovery of Gravitational Radiation." *Physical Review Letters* **22**, 1320–1324.

——— (1984). "Gravitons, Neutrinos and Anti-Neutrinos." *Foundations of Physics* **14**, 1185.

Weber, Joseph, and John Archibald Wheeler (1957). "Reality of the Cylindrical Gravitational Waves of Einstein and Rosen." *Reviews of Modern Physics* **29**, 509–515.

Weisberg, Joel M., and Joseph H. Taylor (1981). "Gravitational Radiation from an Orbiting Pulsar." *General Relativity and Gravitation* **13**, 1–6.

Westpfahl, Konradin (1985). "High Speed Scattering of Charged and Uncharged Particles in General Relativity." *Fortschritte der Physik* **33**, 417–493.

Weyl, Hermann (1921). *Raum, Zeit, Materie*. 4th ed. (Springer, Berlin). Translation by Henry L. Brose as *Space, Time, Matter* (Metheun, London, 1922).

Wheeler, John A. (1961). "Geometrodynamics and the Problem of Motion." *Reviews of Modern Physics* **33**, 63–78.

—— (1962). "Gravitational Quanta and Waves." In *Encyclopedic Dictionary of Physics*, ed. J. Thewlis, p. 506 (Pergamon, London).

—— (1967). "Physics in Transition: Dialogues with Wheeler and Dicke." *Scientific Research* **2**, (May), pp. 50–56.

—— (1998). *Geons, Black Holes and Quantum Foam: A Life in Physics* (Norton, New York).

Wheeler, John Archibald, and Richard Phillips Feynman (1945). "Interaction with the Absorber as the Mechanism of Radiation." *Reviews of Modern Physics* **17**, 157–181.

—— (1949). "Classical Electrodynamics in Terms of Direct Interparticle Action." *Reviews of Modern Physics* **21**, 425–433.

Will, Clifford M. (1993). *Theory and Experiment in Gravitational Physics* (Cambridge University Press, Cambridge).

INDEX

315